D0072488

IMA MONOGRAPH SERIES

IMA MONOGRAPH SERIES

1. H. R. Pitt
   Measure and integration for use

Measure and integration for use

H. R. Pitt, FRS

Former Vice-Chancellor, University of Reading

CLARENDON PRESS · OXFORD
1985

Oxford University Press, Walton Street, Oxford OX2 6DP

London New York Toronto
Delhi Bombay Calcutta Madras Karachi
Kuala Lumpur Singapore Hong Kong Tokyo
Nairobi Dar es Salaam Cape Town
Melbourne Auckland

and associated companies in
Beirut Berlin Ibadan Mexico City Nicosia

Oxford is a trade mark of Oxford University Press

Published in the United States
by Oxford University Press, New York

British Library Cataloguing in Publication Data

Pitt, H. R.
Measure and integration for use.—(IMA monograph series)
1. Measure theory 2. Integrals, Generalized
I. Title II. Series
515.4'3 QA312

ISBN 0-19-853608-9

Set and printed in Northern Ireland by The Universities Press (Belfast) Ltd.

Preface and summary of part one

The purpose of this book, as the title implies, is to provide as simply as possible an account of the Lebesgue theory of measure and integration, with illustrations of its use, which is adequate for practical use in the many fields, both in other branches of mathematics and in the modelling of systems in the real world, for which it is an essential tool. There is no need to labour the point that the theory is useful, for this is self-evident to any practising analyst. What is perhaps less obvious and less widely accepted is the conviction implicit in what follows that the Lebesgue theory is not only more complete and satisfying in itself than the classical Riemann approach, but is also easier to use, and that its practical usefulness more than repays the initial effort of learning it.

Nevertheless, it must be admitted that the theory itself, but not its use, does involve pedagogical difficulties which have long been recognized and which generally arise at a time when a student is overloaded with other new material. The approach to this dilemma which is advocated is radical, but not original, since it is widely practised in other fields even if it has not been highly esteemed among mathematicians. It is to accept that a theory can be used, and its implications understood, without a complete mastery of all the detailed structure of proof on which it rests. This is not to say that such mastery is of little value or that it need never be attempted, but that there is nothing sacrosanct about any traditional order and sequence in the work of learning and using a theory.

This suggests that the reader of this book might find it helpful to concentrate first on learning and understanding the results rather than the detailed proofs of the theorems of Part 1 and follow some of their applications in Part 2 before undertaking a more rigorous study of the earlier work. The argument for this approach is strengthened by the fact that the Lebesgue theory is very much of a piece and has to be available in its entirety before much use can be made of it, and that there is no natural way in which it can be divided up into easy stages. In fact, the only significant break is between the abstract and general outline in Chapter 2 and its particularization in Chapter 3 to Euclidean spaces in which specific algebraic and topological properties are needed to establish the stronger special theorems of the integral calculus. The fact that the abstract generality of the theory in Chapter 2 detracts nothing from its power and utility when applied in the more concrete context of Chapter 3 is a reassuring illustration of the principle of the unity of mathematics and its application to which the IMA is so firmly committed.

The same principle is exemplified also in the later chapters in which three areas of application are selected. The reasons why these were chosen out of the great range of applications of measure theory in modern mathematics are quite obvious. The ideas of area and volume in Chapter 4 are almost inseparable, either logically or historically, from that of measure itself and it would be simply perverse to exclude them. Even at the very practical level of learning and understanding, sets in the plane can be visualized and drawn in a way which provides the most helpful model for the abstract notions of sets, set operations, and set functions needed in Chapter 2. The same might be said of sets in space, at least for those fortunate enough to visualize easily in three dimensions, but not for sets on a line. The difficulty is that linear sets other than intervals are not easy to draw or to disassociate from specific topological properties of the line which are irrelevant to the general case.

There is no need to emphasize the importance of the application of harmonic analysis to physical situations in which wave-forms play a part. Also, apart from this, harmonic analysis is the best example of a branch of mathematics in which it would seem clumsy and unnatural to use a weaker concept of integration. It is enough to think how unsatisfactory a theory it would be without the power and symmetry of the Plancherel and Riesz–Fischer theorems, neither of which is valid outside the Lebesgue theory.

The final chapter is included because a comprehensive theory of probability is impossible without a theory of measure—for probability *is* measure—and the subject is crippled if it lacks the potentiality for limit operations which the Lebesgue theory, quite characteristically, provides.

The following summary contains nothing which is not in the main text, but it may be helpful as a guide to the structure of the theory and an indication of the more significant steps. Part 2 is not so compact and coherent and it is not possible to summarize it in the same way.

The general approach, and some of the material, is similar to that of an earlier book by the author (*Integration, measure and probability*, Oliver and Boyd, London, 1968) which is now out of print.

Summary of Part one

The primary concepts in measure theory, after the familiar notions of sets, set operations, and set functions, are additive and completely additive set functions over a ring or a σ-ring. A non-negative and completely additive real-valued function over a σ-ring is called a measure and the significance of the Hahn–Jordan decomposition theorem (Theorem 3 of

Chapter 1) is that it reduces the theory of general completely additive set functions to that of measure. The development of the theory in Euclidean space $\mathcal{R}^k$ follows from properties of the ring of *figures* consisting of finite unions of rectangular sets and the crucial fact that volume is a completely additive function on this ring. In more abstract spaces, all that is needed in place of figures and their volume is a ring of analogous *simple sets* I on which a completely additive non-negative set function μ with values $\mu(I)$ is defined. This may be called a simple measure and the essence of Chapter 2 is that its domain can be extended to a σ-ring on which it becomes a true and complete measure and that an integral $\int f$ can be defined over a wide domain of integrable functions f. It turns out, in fact, to be quicker and more economical to derive the measure from the integral rather than the other way round as in the original development of Lebesgue and his contemporaries.

The key to the development is the use of simple functions θ and the metric $\|f\|$ which is defined, if we allow positive infinite values, for all real valued functions f and provides an indication of the 'overall size' of f. In particular, a null function for which $\|f\|=0$, can be neglected for all practical purposes and condition $\|f\|<\infty$ is found to be of crucial significance throughout. In short, a function f with positive and negative components f^+ and f^- is integrable if a sequence of simple functions θ_n can be defined so that $\|f-\theta_n\|\to 0$, and its integral is defined by $\int f=\|f^+\|-\|f^-\|$. In particular, a null set X is one whose characteristic function γ taking values 1 in X and 0 outside X is null; and a property is said to hold almost everywhere (a.e.) if it holds everywhere outside a null set.

Many properties of the integral follow very easily from this, but it is important to remember that although they look familiar, and are certainly true in ordinary Euclidean spaces, they are valid in completely abstract spaces $\mathcal{X}$ about which nothing whatever is assumed except the existence of a simple measure. This contrasts very sharply with the classical Riemann theory in which the specific topological properties of $\mathcal{R}^k$ are required from the start. The result which most clearly shows the power and utility of the Lebesgue theory, both in the abstract context of Chapter 2 and in the way it is characteristic of the theory in $\mathcal{R}^k$, is the celebrated convergence theorem (Theorem 18 of Chapter 2) of Lebesgue. This says, very simply, that $\int f_n\to\int f$ if $f_n\to f$ a.e. provided only that $|f_n|\leqslant\lambda$, $\|\lambda\|<\infty$. In other words, $\lim\int f_n=\int\lim f_n$, so that the order of taking the limit and the integral can be reversed, provided that the functions f_n are 'not too big'; and there is no need for uniformity of convergence or any other condition.

The ideas of measure and measurability follow very quickly once the integral has been established, for a function is measurable if it is the limit a.e. of simple functions, and a set is measurable if its characteristic function is measurable. The integral of such a characteristic function then defines the value of a non-negative completely additive set function on the σ-ring of measurable sets and therefore gives a measure which is an extension of the original simple measure.

Since the abstract space $\mathscr{X}$ is not assumed to have a topology, there may not be associated with it any notion of a derivative in the traditional sense of the differential calculus in $\mathscr{R}^k$. There remains, nevertheless, in the abstract Lebesgue theory a process of differentiation of a set function which can be regarded quite precisely as an inverse of the operation of integration. It is dependent on the definition of an absolutely continuous completely additive set function F, with values $F(X)$, as one for which $F(X) \to 0$ uniformly as $\mu(X) \to 0$ and it is important to note that, in spite of the use of the word continuous, the property is purely measure-theoretical and need have no topological connotation. The Radon–Nikodym theorem (Theorem 28 of Chapter 2) shows that an integral $F(X)$ of an integrable function f over the set X is the value of an absolutely continuous set function F and that, conversely, any such set function is the integral of a unique function f called its Radon–Nikodym derivative. The theorem can be extended to characterize set functions which are not absolutely continuous by expressing them as sums of absolutely continuous and residual (or singular) components, and it then provides a very powerful tool for the establishment in a simple and direct way of Stieltjes integrals and the most general formula for the change of variable in Lebesgue integrals.

It is shown in Section 2.5 that measures μ and v in two spaces $\mathscr{X}$ and $\mathscr{Y}$ define in a very natural way a product measure m in their product space $\mathscr{X} \times \mathscr{Y}$. The basic theorem of Fubini (Theorem 33) then asserts the existence and equality of the integrals $\int f \, dm$, $\int f \, d\mu \int f \, dv$ and $\int dv \int f \, d\mu$ provided only that any one of them exists with $|f|$ in place of f. In other words, the order of the repeated operations of integration with respect to μ and v can be interchanged or the repeated integration replaced by a single integration with respect to their product measure provided only that f is 'not too big'. The analogy with Lebesgue's convergence theorem is obvious since both involve the commutativity of operations and both display great merits over the corresponding results for the Riemann integral in which heavy continuity or uniformity conditions are required. The final Section 2.6 shows that familiar and useful inequalities can be established as easily in general measure spaces as in $\mathscr{R}^k$.

The whole of the theory in Chapter 2 applies immediately, of course to the familiar measure and integral in $\mathscr{R}^k$ based on figures as the simple sets and volume as the simple measure, and it therefore constitutes a major contribution to the integral calculus in $\mathscr{R}^k$. Chapter 3 completes this by adding the results which depend on the specific properties of $\mathscr{R}^k$

and which may be false, or even meaningless in the more abstract context of Chapter 2. Most of this is straightforward and familiar and confirms the simplicity and power of the Lebesgue integral as a working tool as illustrated, for example, in Theorem 9 and the simple condition for the interchange of the order of integration with respect to one variable and differentiation with respect to another.

But the central result, the fundamental theorem of calculus, is not easy or straightforward, and a little thought is needed to appreciate its full significance. This depends on the fact that in $\mathscr{R}$ we have two quite distinct ways of defining a derivative. The first is by the Radon–Nikodym theorem of Chapter 2; the second is by the traditional process of differentiation. It is by no means obvious why the two processes should lead to the same result, but Theorem 11 asserts that this is so, and for that we must be profoundly thankful!

The fundamental theorem can be extended from $\mathscr{R}$ to $\mathscr{R}^k$ without difficulty and then provides by far the most direct treatment of the otherwise difficult and tedious problem of change of variable in multiple integrals.

June 1984 H. R. P.
Reading

Contents

PART I

PART II

PART I

1
Sets and set-functions

1.1 Set operations

The modern theories of measure and probability are based on fundamental notions of sets and set-functions which are now familiar and readily accessible. It is therefore simply on grounds of convenience that we begin with a brief summary of these ideas.

We shall be concerned with a space $\mathscr{X}$ of elements (points) x, subsets X, X_1, X_2, of $\mathscr{X}$, and certain systems of subsets. In accordance with common usage, we write $x \in X$ to mean that the point x belongs to the set X and $X_1 \subset X_2$ to mean that the set X_1 is included in (or is a subset of) the set X_2. The empty set is denoted by $\varnothing$, and it is plain that $X \subset X$ and $\varnothing \subset X \subset \mathscr{X}$. Moreover, $X_1 \subset X_2$ and $X_2 \subset X_3$ together imply that $X_1 \subset X_3$, while $X_1 \subset X_2$ and $X_2 \subset X_1$ imply that $X_1 = X_2$.

The **operations** on sets are defined as follows. The set of points of $\mathscr{X}$ which are not in X is called the **complement** of X and written X'. It is clear that $(X')' = X$, $\varnothing' = \mathscr{X}$, $\mathscr{X}' = \varnothing$ and, if $X_1 \subset X_2$, then $X_2' \subset X_1'$.

The **union** of any collection of sets is the set of points which belong to at least one of them. The collection need not be finite or even countable. The union of two sets X_1, X_2 is written $X_1 \cup X_2$. The union of a finite or countable collection of sets X_n $(n = 1, 2, \ldots)$ is written $\bigcup_n X_n$. It is plain from the definition that the union operation does not depend on any particular ordering of the component sets and that there is no limit process involved even when the number of terms is infinite.

The **intersection** of a collection of sets is the set of points which belong to every set of the collection. The intersection of two sets is written $X_1 \cap X_2$, while the intersection of a sequence X_n is written $\bigcap_n X_n$. A pair of sets is **disjoint** if their intersection is $\varnothing$ and a collection of sets is disjoint if every pair is disjoint. In particular, the union of a sequence of sets $X_1, X_2, \ldots$ can be expressed on the union of the disjoint sets X_1, $X_1' \cap X_2$, $X_1' \cap X_2' \cap X_3, \ldots$

The **difference** $X_1 - X_2 = X_1 \cap X_2'$ between the sets X_1 and X_2 is the set of points of X_1 which do not belong to X_2.

It is clear that each of the union and intersection operations is commutative and associative, while each is distributive with respect to the other in the sense that

$$X \cap (X_1 \cup X_2) = (X \cap X_1) \cup (X \cap X_2),$$
$$X \cup (X_1 \cap X_2) = (X \cup X_1) \cap (X \cup X_2)$$

or, more generally,

$$X \cap (\bigcup X_v) = \bigcup (X \cap X_v), \qquad X \cup (\bigcap X_v) = \bigcup (X \cap X_v).$$

Furthermore, the operations are related to complementation by

$$X \cap X' = \varnothing, \qquad X \cup X' = \mathscr{X}, \qquad (X_1 \cup X_2)' = X_1' \cap X_2',$$
$$(X_1 \cap X_2)' = X_1' \cup X_2'$$

and, more generally,

$$(\bigcup X)' = \bigcap x', (\bigcap X)' = \bigcup X'.$$

The sequence of sets X_n is **increasing** (and we write $X_n \uparrow$) if, for each n, $X_n \subset X_{n+1}$, and **decreasing** ($X_n \downarrow$) if $X_{n+1} \subset X_n$. The **upper limit,** $\limsup X_n$, of a sequence is the set of points belonging to infinitely many of the sets; the **lower limit,** $\liminf X_n$, is the set of points belonging to X_n for all but a finite number of values of n. It follows that $\liminf X_n \subset \limsup X_n$. If

$$\limsup X_n = \liminf X_n = X,$$

X is called the **limit** of X_n and X_n is said to **converge** to X. We then write $X_n \to X$, or $X_n \uparrow X$, $X_n \downarrow X$ in the cases when X_n decreases or increases. It is plain from the definitions that

$$\liminf X_n = \bigcup_{n=1}^{\infty} \bigcap_{m=n}^{\infty} X_m, \qquad \limsup X_n = \bigcap_{m=1}^{\infty} \bigcup_{n=m}^{\infty} X_n.$$

In particular, if X_n decreases,

$$\bigcap_{m=n}^{\infty} X_m = \bigcap_{m=1}^{\infty} X_m, \qquad \bigcup_{n=m}^{\infty} X_n = X_m,$$

$$\liminf X_n = \bigcap_{m=1}^{\infty} X_m = \limsup X_n, \qquad \lim X_n = \bigcap_{m=1}^{\infty} X_m.$$

Similarly, if X_n increases, $\lim X_n = \bigcup_{n=1}^{\infty} X_n$.

1.2 Additive systems of sets

A non-empty collection $\mathbf{S}$ of sets X in $\mathscr{X}$ is called a **ring** if, whenever X_1 and X_2 belong to $\mathbf{S}$, so do $X_1 \cup X_2$ and $X_1 - X_2$. The empty set can be expressed as $\varnothing = X - X$ and therefore belongs to every ring. Moreover, since

$$X_1 \cap X_2 = (X_1 \cup X_2) - \{(X_1 - X_2) \cup (X_2 - X_1)\},$$

the intersection of two sets of $\mathbf{S}$ also belongs to $\mathbf{S}$. The result of applying a finite number of union, intersection, or difference operations to elements of a ring is therefore to give another element of the ring. In other words, the ring is closed under these operations.

The ring is also closed under the complement operation if (and only if) $\mathscr{X}$ itself belongs to the ring, but we do not assume that this is the case in general.

A ring **S** which contains $\mathscr{X}$ and has the further property that the union of any countably infinite collection of its members also belongs to **S** is called a **σ-ring**. Since

$$\bigcap X_n = (\bigcup X_n')',$$

it follows that a σ-ring also contains the intersection of any countable collection of its members and is thus closed under union, intersection, difference, and complement operations repeated a finite or countably infinite number of times.

We shall often be concerned with the construction of a ring or σ-ring to include a given collection **T** of sets of $\mathscr{X}$ and the following theorem is fundamental.

Theorem 1. *A given collection* **T** *of subsets of space* $\mathscr{X}$ *is contained in a unique minimal ring (σ-ring) which is contained in every ring (σ-ring) which contains* **T**.

The minimal ring is called the ring **generated** by **T**. The minimal σ-ring generated by **T** is called the **Borel extension** of **T**, and its members are called **Borel sets**.

There is at least one ring containing **T**, namely the ring of all subsets of $\mathscr{X}$. We consider all such rings. It is plain that their intersection, consisting of the sets which belong to every such ring is itself a ring and has the properties stated, and the proof remains valid if we substitue σ-ring for ring throughout. In applications, **T** is usually a ring but not a σ-ring. The theorem stills holds in a trivial sense if **T** is itself a σ-ring, but it is then its own Borel extension.

If $\mathscr{X}$ is the space $\mathscr{R}$ of real numbers, the intervals do not form a ring since the union of two intervals need not be an interval. However, the collection of sets which consist of a finite union of intervals of type $a \leqslant x < b$ do form a ring which is plainly the ring generated by these intervals in the sense of the last theorem. This ring does not contain $\mathscr{R}$ and is not a σ-ring. There is a straightforward generalization to the space $\mathscr{R}_k$ of real vectors $(x_1, x_2 \dots x_k)$ in which a finite union of rectangles $a_j \leqslant x_j < b_j$ $(j = 1, 2, \dots k)$ is called a **figure** and the ring generated by the rectangles is the **ring** of **figures**.

The ring of figures, particularly in the cases $k = 1$, 2, and 3, is of fundamental importance in the theory of measure and integration. The ring is easy to visualize and provides a useful concrete example on which the significance of abstract theorems may be illustrated.

1.3. Additive set functions

A set function μ is a function whose range of definition is a system (usually a ring) of sets X and whose values $\mu(X)$ belong to some appropriate space. We shall deal in this book only with set functions whose values are real numbers, but it is convenient to augment these by appending the elements $\pm\infty$ and giving them algebraic and order properties in relation to real numbers ξ according to the following scheme.

$$-\infty < \xi < \infty, \qquad (\pm\infty)+(\pm\infty)=\xi+(\pm\infty)=\pm\infty;$$

$$\xi(\pm\infty)=(\pm\infty)\xi=\pm\infty \quad \text{or} \quad \mp\infty \quad \text{according} \quad \text{as } \xi>0, \xi<0.$$

$$(\pm\infty)(\pm\infty)=\infty, \qquad (\pm\infty)(\mp\infty)=-\infty.$$

The operations $\infty-\infty$, ∞/∞, $0.\infty$ are not defined.

A set function μ defined in a ring $\mathbf{S}$ is called **additive** if

$$\mu\left(\bigcup X_n\right)=\sum \mu(X_n)$$

for every *finite* disjoint collection of sets in $\mathbf{S}$. The function is **completely additive** in $\mathbf{S}$ if the additivity property holds also for countably infinite collections of sets X_n in $\mathbf{S}$ provided also that $\bigcup X_n$ belongs to $\mathbf{S}$. This last proviso is not needed when $\mathbf{S}$ is a σ-ring since it is satisfied automatically. Otherwise, it must be retained and only certain sequences of sets (those whose union belongs to $\mathbf{S}$) may be admitted.

We shall always assume that $\mu(X)$ is finite for at least one set X, so that $\mu(X)=\mu(X\cup\emptyset)=\mu(X)+\mu(\emptyset)$ and $\mu(\emptyset)=0$. It is not possible for an additive set function to take the value $+\infty$ on one set and $-\infty$ an another, since its value on their union would then be undefined; and we shall exclude this possibility by admitting $+\infty$ as a possible value, but not $-\infty$. A set function which takes neither of the values $\pm\infty$ is called **finite**.

A non-negative and completely additive function μ defined over a σ-ring is called a **measure**. It is called a **probability measure** if $\mu(\mathscr{X})=1$.

A set function μ is said to be **continuous from below** at X if $\mu(X_n)\to\mu(X)$ whenever $X_n\uparrow X$. It is **continuous from above** at X if $\mu(X_n)\to\mu(X)$ whenever $X_n\downarrow X$ and $\mu(X_N)<\infty$ for some N. It is **continuous** at X if it is continuous from above and below at X unless $X=\emptyset$, when continuity means continuity from above. The relationship between additivity and complete additivity is expressed in terms of continuity in the following theorem.

Theorem 2. *A completely additive function μ is continuous. Conversely, an* *additive function is completely additive if it is continuous from below at every set or if it is finite and continuous at* $\varnothing$.

(The ring in which the function is defined may, but need not, be a σ-ring.)

Suppose that μ is completely additive. If $X_n \uparrow X$ and $\mu(X_n) \neq \pm\infty$ for every n, we have

$$X = X_1 \cup \bigcup_{v=2}^{\infty} (X_v - X_{v-1}),$$

$$\begin{aligned}\mu(X) &= \mu(X_1) + \sum_{v=2}^{\infty} \mu(X_v - X_{v-1}) \\ &= \mu(X_1) + \lim_{n\to\infty} \sum_{v=2}^{n} \mu(X_v - X_{v-1}) = \lim_{n\to\infty} \mu(X_n).\end{aligned}$$

If $\mu(X_M) = \infty$, we have $\mu(X_n) = \mu(X_M) + \mu(X_n - X_M) = \infty$ and $\mu(X) = \mu(X_M) + \mu(X - X_M) = \infty$, since $\mu(X) \neq -\infty$ for any X. On the other hand, if $X_n \downarrow X$ and $\mu(X_N) < \infty$, we have

$$X_N = X \cup \bigcup_{v=N}^{\infty} (X_v - X_{v+1}),$$

$$\begin{aligned}\mu(X_N) &= \mu(X) + \sum_{v=N}^{\infty} \mu(X_v - X_{v+1}) = \mu(X) + \lim_{n\to\infty} \sum_{v=N}^{n} \mu(X_v - X_{v+1}) \\ &= \mu(X) + \mu(X_N) - \lim_{n\to\infty} \mu(X_n).\end{aligned}$$

Hence, μ is continuous.

To prove the converse, we suppose that Y_v are disjoint and

$$X = \bigcup_{v=1}^{\infty} Y_v, \qquad X_n = \bigcup_{v=1}^{n} Y_v,$$

so that $X_n \uparrow X$. Then

$$\sum_{v=1}^{\infty} \mu(Y_v) = \lim_{n\to\infty} \sum_{v=1}^{n} \mu(Y_v) = \lim_{n\to\infty} \mu(X_n)$$

by finite additivity, and this is $\mu(X)$ if we assume that μ is continuous below at X. But we may also write

$$\mu(X) = \mu(X_n) + \mu(X - X_n) = \sum_{v=1}^{n} \mu(Y_v) + \mu(X - X_n)$$

by finite additivity, and this gives the same conclusion if μ is finite and continuous at $\varnothing$.

The next theorem of Hahn and Jordan shows how a set function taking positive and negative values can be expressed as the sum of positive and negative component set functions. This is of great practical importance, since it means that results about general set functions can be deduced almost immediately from those about measures.

Theorem 3. (*Hahn–Jordan decomposition theorem*). *Suppose that μ is completely additive in a σ-ring* **S** *of sets X in $\mathscr{X}$, let M, m be the upper and lower bounds of $\mu(X)$ in* **S** *and $-\infty < m \leq M \leq \infty$. Then $\mathscr{X} = \mathscr{X}^+ \cup \mathscr{X}^-$, where $\mathscr{X}^+$ and $\mathscr{X}^-$ belong to* **S** *(and either may be empty) and*

$$m = \mu(\mathscr{X}^-) \leq \mu(X) \leq 0 \quad \text{for} \quad X \subset \mathscr{X}^+, \tag{1}$$

$$0 \leq \mu(X) \leq \mu(\mathscr{X}^+) = M \quad \text{for} \quad X \subset \mathscr{X}^+. \tag{2}$$

Moreover, $\mu(X) = \mu^+(X) + \mu^-(X)$, where $\mu^+(X)$, and $\mu^-(X)$ are uniquely defined and completely additive in **S** *and*

$$\mu^+(X) = \mu(X \cap \mathscr{X}^+) = \sup_{Y \subset X} \mu(Y) \geq 0,$$

$$\mu^-(X) = \mu(X \cap \mathscr{X}^-) = \inf_{Y \subset X} \mu(Y) \leq 0.$$

Corollary 1. *If $\mu(X) < \infty$ for all X, then $M < \infty$ and $\mu(X)$ is bounded.*

Corollary 2. *If $|\mu|(X) = \mu^+(X) - \mu^-(X)$, then*

$$\sup_{Y \subset X} |\mu(Y)| \leq |\mu|(X).$$

Let $\varepsilon_k > 0$, $\sum \varepsilon_k < \infty$. Then we can define sets A_k in **S** so that $m \leq \mu(A_k) \leq m + \varepsilon_k$ for $k = 1, 2, 3, \ldots$.

Since

$$A_2 = (A_1 \cap A_2) \cup (A_2 - A_1), \qquad (A_2 - A_1) \cup A_1 = A_1 \cup A_2,$$

we have

$$\mu(A_2) = \mu(A_1 \cap A_2) + \mu(A_2 - A_1), \qquad \mu(A_2 - A_1) + \mu(A_1) = \mu(A_1 \cup A_2),$$

$$m \leq \mu(A_1 \cap A_2) = \mu(A_2) + \mu(A_1) - \mu(A_1 \cup A_2)$$

$$\leq m + \varepsilon_1 + m + \varepsilon_2 - m = m + \varepsilon_1 + \varepsilon_2.$$

The same argument applied to the sets $A_n, A_{n+1}, \ldots A_p$ gives

$$m \leq \mu\left\{\bigcap_{v=n}^{p} A_v\right\} \leq m + \sum_{v=n}^{p} \varepsilon_v,$$

and it follows from the continuity of μ that

$$m \leq \mu\left\{\bigcap_{v=n}^{\infty} A_v\right\} \leq m + \sum_{v=n}^{\infty} \varepsilon_v.$$

If we now define $\mathscr{X}^- = \liminf A_k$, we have

$$\bigcap_{v=n}^{\infty} A_v \uparrow \mathscr{X}^- \quad \text{as} \quad n \to \infty,$$

and it follows again from the continuity of μ that

$$\mu(\mathscr{X}^-) = \lim_{n\to\infty} \mu\left\{\bigcap_{v=n}^{\infty} A_v\right\} = m.$$

If $X \subset \mathscr{X}^-$, we have $\mathscr{X}^- = X \cup (\mathscr{X}^- - X)$.

$$m = \mu(\mathscr{X}^-) = \mu(X) + \mu(\mathscr{X}^- - X) \geqslant \mu(X) + m,$$

and so $\mu(X) \leqslant 0$, which completes the proof of (1). On the other hand, if $X \subset \mathscr{X}^+ = \mathscr{X} - \mathscr{X}^-$, we have

$$m \leqslant \mu(X \cup \mathscr{X}^-) = \mu(X) + \mu(\mathscr{X}^-) = \mu(X) + m, \qquad \mu(X) \geqslant 0.$$

We can now define sets B_n of $\mathbf{S}$ so that $\mu(B_n) \to M$, and since $B_n = (B_n \cap \mathscr{X}^-) \cup (B_n \cap \mathscr{X}^+)$, it follows that

$$\begin{aligned}\mu(\mathscr{X}^+) &= \mu(B_n \cap \mathscr{X}^+) + \mu(\mathscr{X}^+ - B_n) \geqslant \mu(B_n \cap \mathscr{X}^+) \\ &\geqslant \mu(B_n \cap \mathscr{X}^-) + \mu(B_n \cap \mathscr{X}^+) = \mu(B_n)\end{aligned}$$

and $\mu(\mathscr{X}^+) = M$.

Finally, if $Y \subset X$,

$$\begin{aligned}\mu(Y) &= \mu(Y \cap \mathscr{X}^+) + \mu(Y \cap \mathscr{X}^-) \leqslant \mu(Y \cap \mathscr{X}^+) \\ &\leqslant \mu(Y \cap \mathscr{X}^+) + \mu\{(X - Y) \cap \mathscr{X}^+\} = \mu(X \cap \mathscr{X}^+)\end{aligned}$$

and $\mu(Y) \geqslant \mu(X \cap \mathscr{X}^-)$ similarly. The complete additivity of $\mu^+(X)$ and $\mu^-(X)$ is obvious.

1.4 Additive functions on rings of figures in $\mathcal{R}_k$

We introduce here certain important set functions defined on the ring of figures described in Section 1.2. We begin with the case $k = 1$ and suppose that μ is a non-decreasing and finite point function in $-\infty < x < \infty$. It is familiar that the left and right limits $\mu(x-0)$, $\mu(x+0)$ exist at every point x and that $\mu(x-0) \leqslant \mu(x) \leqslant \mu(x+0)$. We then define

$$\mu(I) = \sum \{\mu(b_v - 0) - \mu(a_v - 0)\} \tag{1}$$

when I is the figure formed by the union of the disjoint intervals $a_v \leqslant x < b_v$. It is easy to verify that $\mu(I)$ is non-negative and finitely additive in the ring of figures I. In the important special case $\mu(x) = x$, $\mu(I)$ is simply the length of I.

The following result, that μ is *completely* additive, is of fundamental importance in the theory of integration in $\mathcal{R}$.

Theorem 4. *The set function defined by* (1) *in respect to a non-decreasing point function* μ *is completely additive in the ring of figures in* $\mathcal{R}$.

After Theorem 2, it is sufficient to show that if I_n are figures and $I_n \downarrow 0$, then $\mu(I_n) \to 0$. If not, we can define $d > 0$ so that $\mu(I_n) > d$ for all n. But since $\mu(b-0) = \lim_{x \to b-0} \mu(x)$, we can construct figures I_n^* and closed figures J_n so that

$$I_n^* \subset J_n \subset I_n, \qquad \mu(I_n - I_n^*) < d/2^{n+1}.$$

Then

$$\mu\left\{\bigcap_{\upsilon=1}^{n} I_\upsilon^*\right\} = \mu\left\{\bigcap_{\upsilon=1}^{n} [I_\upsilon - (I_\upsilon - I_\upsilon^*)]\right\} \geq \mu(I_n) - \sum_{\upsilon=1}^{n} \mu(I_\upsilon - I_\upsilon^*) > d/2.$$

The sets $\bigcup_{\upsilon=1}^{n} J_\upsilon$ are therefore closed, decreasing and not empty and we appeal to the theorem that the intersection of such a sequence is not empty to deduce that J_n, and therefore I_n, have a common point. This is inconsistent with our assumption that $I_n \downarrow 0$, and it follows that $\mu(I_n) \to 0$.

If μ is not necessarily monotonic, but

$$\sum |\mu(\beta_\upsilon) - \mu(\alpha_\upsilon)|$$

is bounded for all finite sets of non-overlapping intervals $(\alpha_\upsilon, \beta_\upsilon)$ in an interval $I(a \leq x < b)$, we say that $\mu(x)$ has **bounded variation,** in I, and the result of Theorem 4 can be extended easily to point functions of this kind. The function μ, defined by

$$\mu_1(x) = \sup \sum \{\mu(\beta_\upsilon) - \mu(\alpha_\upsilon)\},$$

where the summation is over all finite sets of disjoint intervals $(\alpha_\upsilon, \beta_\upsilon)$ in $a \leq \alpha_\upsilon < \beta_\upsilon < x$, is plainly increasing and bounded in I, and it is easy to see that $\mu_2 = \mu_1 - \mu$ has the same property. Thus we can express μ as the difference $\mu_1 - \mu_2$ of two bounded, increasing point functions. These define completely additive set functions with values $\mu_1(I)$, $\mu_2(I)$ over the ring of figures, by Theorem 4, and so $\mu = \mu_1 - \mu_2$ is also completely additive and $\mu(I)$ satisfies (1).

This shows that there is no gain in generality in considering any but non-negative functions μ and we shall generally make this restriction throughout the book except in a few places where the contrary is stated.

If μ is of bounded variation in every finite interval, it can still be expressed in the form $\mu_1 - \mu_2$ in which μ_1, μ_2 will be increasing, but not necessarily bounded. They are bounded in $(-\infty, \infty)$ if and only if

$$\sum |\mu(\beta_v) - \mu(\alpha_v)|$$

is bounded for all non-overlapping intervals (α_v, β_v) in $(-\infty, \infty)$, and μ is then said to have bounded variation in $(-\infty, \infty)$.

This result can be generalized, with hardly more than verbal change in the proof, to rings of figures in $\mathcal{R}^k$ $(k>1)$, but the close relationship between the set function on the ring and a point function over $\mathcal{R}^k$ is then not so important. We shall not have occasion to use anything beyond the familiar case in which the set function extends the concept of area or volume.

The volume (area if $k=2$) of the k-dimensional rectangle $a_j \leqslant x_j < b_j$ is defined by

$$\prod_{j=1}^{k} (b_j - a_j)$$

and it is easy to verify that if a figure is decomposed into rectangles in different ways, the sum of volumes of its component rectangles is still the same. This sum $\mu(I)$ for a figure I can therefore be called the volume of I without ambiguity, and the following theorem is a straightforward generalization of the case $F(x) = x$ of Theorem 4.

Theorem 5. *The volume $\mu(I)$ of a figure I is completely additive in the ring of figures in $\mathcal{R}^k$.*

2
General theory of integration and measure

2.1 Definition of an integral

We suppose that the space $\mathscr{X}$ contains a ring of sets, called **simple sets,** on which a completely additive, finite-valued and non-negative set function $\mu(I)$ is defined, and that $\mathscr{X}$ is the limit of a sequence of simple sets. We call such a set function a **simple measure.** We then consider functions f defined on $\mathscr{X}$ and taking real values (and possibly $\pm\infty$), and use the customary definitions

$$f^{+}(x)=\sup\{f(x), 0\}, \qquad f^{-}(x)=\inf\{f(x), 0\} \tag{1}$$

so that

$$f=f^{+}+f^{-}, \qquad |f|=f^{+}-f^{-}.$$

The theory can be extended to complex-valued functions by treating their real and imaginary parts separately and using Theorem 11 and the inequalities $|\alpha|,\ |\beta|\leqslant|\alpha+i\beta|\leqslant|\alpha|+|\beta|$ whenever necessary. We write $f\geqslant g$ if $f(x)\geqslant g(x)$ for every x in $\mathscr{X}$, and $f_n\to f$ if $f_n(x)\to f(x)$ for every x in $\mathscr{X}$ as $n\to\infty$. We also need a general rule that the values of functions must be restricted, whenever necessary, to ensure the validity of any algebraic operations done on them. For example, $f\pm g$ cannot be defined if $f=\infty$, $g=\mp\infty$; fg cannot be defined if $f=\infty$, $g=0$.

A function θ is called a **simple function** if it takes constant finite values a_j in each of a finite number of simple sets I_j and vanishes elsewhere, and we define

$$A(\theta)=\sum a_j\mu(I_j).$$

It is plain that if θ is simple, so are θ^{+}, θ^{-}, and $|\theta|$, and that any finite linear combination $\sum a_v\theta_v$ of simple functions is also simple with

$$A\left(\sum a_v\theta_v\right)=\sum a_vA(\theta_v). \tag{2}$$

In particular,

$$A(\theta)=A(\theta^{+})+A(\theta^{-}), \qquad A(|\theta|)=A(\theta^{+})-A(\theta^{-}).$$

Finally,

$$A(\theta)\geqslant 0 \quad \text{if} \quad \theta\geqslant 0,$$
$$A(\theta_1)\geqslant A(\theta_2) \quad \text{if} \quad \theta_1\geqslant\theta_2.$$

For any function f (not generally simple), we define

$$\|f\|=\inf\sum A(\theta_m)$$

for all sequences of simple functions θ_m which satisfy

$$\theta_m\geqslant 0, \qquad |f|\leqslant\sum\theta_m.$$

Such a sequence always exists since we may define simple sets I_m so that $I_m\to\mathscr{X}$ and define $\theta_m=m$ in I_m, $\theta_m=0$ in I'_m. We write $\|f\|=\infty$ if $\sum A(\theta_m)$ diverges for every such sequence θ_n, but we note that $\|f\|$ may be finite even if f takes infinite values at certain points. It is obvious that $\||f|\|=\|f\|$, $\|af\|=|a|\,\|f\|$, $\|f^+\|\leqslant\|f\|$, $\|f^-\|\leqslant\|f\|$.

Theorem 1. *If* $|f|\leqslant|g|$, *then* $\|f\|\leqslant\|g\|$.

If $\varepsilon>0$, we can define simple functions $\theta_m\geqslant 0$ so that $|f|\leqslant|g|\leqslant\sum\theta_m$, while $\sum A(\theta_m)\leqslant\|g\|+\varepsilon$. Then $\|f\|\leqslant\sum A(\theta_m)\leqslant\|g\|+\varepsilon$ for every positive ε and therefore $\|f\|\leqslant\|g\|$.

Theorem 2. *If* $\sum f_n$ *is defined for a finite or countable sequence of functions* f_n, *then* $\|\sum f_n\|\leqslant\sum\|f_n\|$.

If $\varepsilon>0$, we choose simple functions $\theta_{nm}\geqslant 0$ so that $|f_n|\leqslant\sum_m\theta_{nm}$, $\sum_m A(\theta_{nm})\leqslant\|f_n\|+2^{-n}\varepsilon$. Then

$$|\sum f_n|\leqslant\sum_n\sum_m\theta_{nm},$$

$$\|\sum f_n\|\leqslant\sum_n\sum_m A(\theta_{nm})\leqslant\sum_n\{\|f_n\|+2^{-n}\varepsilon\}\leqslant\sum\|f_n\|+\varepsilon,$$

which is sufficient.

This last result for two functions f_1, f_2 shows that the functions f for which $\|f\|<\infty$ form a linear metric space with $\|f\|$ as metric.

Theorem 3. *If* $\|f_n\|<\infty$ *and* $\|f-f_n\|\to 0$, *then* $\|f\|$, $\|f^+\|$, $\|f^-\|$ *are finite and* $\|f^+-f_n^+\|\to 0$, $\|f^--f_n^-\|\to 0$, $\||f|-|f_n|\|\to 0$, $\|f_n\|\to\|f\|$, $\|f_n^+\|\to\|f_n^+\|$, $\|f_n^-\|\to\|f^-\|$, $\||f_n|\|\to\||f|\|$.

The first part follows from the inequalities $\|f\|=\|f_n+f-f_n\|\leqslant\|f_n\|+\|f-f_n\|<\infty$, by Theorem 2. The second part follows from Theorem 1 and the elementary inequalities

$$|f^+-f_n^+|\leqslant|f-f_n|, \qquad |f^--f_n^-|\leqslant|f-f_n|, \qquad ||f|-|f_n||\leqslant|f-f_n|.$$

For the last part, we observe that

$$\|f\|-\|f_n-f\|\leqslant\|f_n\|\leqslant\|f\|+\|f_n-f\|,$$

by Theorem 2.

Theorem 4. *If θ_n is a sequence of simple functions and $\theta_n\downarrow 0$ for every x, then $A(\theta_n)\to 0$.*

Let $C=\sup\theta_1(X)$ and let I_1 be the (simple) set in which $\theta_1(x)>0$. Then if $\varepsilon>0$, the set I_n in which $\theta_n(X)\geqslant\varepsilon$ is simple and

$$A(\theta_n)\leqslant C\mu(I_n)+\varepsilon\mu(I_1).$$

Since $I_n\downarrow 0$ and μ is completely additive, it follows from Theorem 2 of Chapter 1 that $\mu(I_n)\to 0$ and so $\limsup A(\theta_n)\leqslant\varepsilon\mu(I_1)$ for every $\varepsilon>0$, and therefore $A(\theta_n)\to 0$.

Theorem 5. *If θ is a simple function,*

$$\|\theta\|=A(|\theta|)=A(\theta^+)-A(\theta^-)=\|\theta^+\|+\|\theta^-\|,$$
$$A(\theta)=A(\theta^+)+A(\theta^-)=\|\theta^+\|-\|\theta^-\|.$$

In particular,

$$\|\theta^+\|=A(\theta^+),\ \|\theta^-\|=-A(\theta^-)\ \textit{and if}\ \theta\geqslant 0,\ \|\theta\|=A(\theta).$$

It is obviously enough to consider the case $\theta\geqslant 0$. It follows from the trivial observation $\theta\leqslant\theta$ that $\|\theta\|\leqslant A(\theta)$. Now suppose that $\theta\leqslant\sum\theta_m$, $\theta_m\geqslant 0$, and let

$$\eta_M=\theta-\sum_{m=1}^{M}\theta_m.$$

Then η_M is simple and decreases and $\lim\eta_M\leqslant 0$ as $M\to\infty$. It follows that $\eta_M^+\downarrow 0$ and therefore, by Theorem 4, $A(\eta_M^+)\to 0$. But

$$\theta=\sum_{m=1}^{M}\theta_m+\eta_M\leqslant\sum_{m=1}^{M}\theta_m+\eta_M^+,$$

$$A(\theta)\leqslant\sum_{m=1}^{M}A(\theta_m)+A(\eta_M^+),$$

and on letting $M\to\infty$, we have $A(\theta)\leqslant\sum A(\theta_m)$. Since this holds for all sequences θ_m with $\theta_m\geqslant 0$, $\theta\leqslant\sum\theta_m$, it follows that $A(\theta)\leqslant\|\theta\|$.

We say that f is **integrable** in $\mathscr{X}$ with respect to μ if we can define a sequence of simple functions θ_n so that $\|f-\theta_n\|\to 0$. If this is satisfied, it follows from Theorem 3 that $\|f\|$, $\|f^+\|$, $\|f^-\|$ are finite, and we write

$$\int f\,\mathrm{d}\mu=\|f^+\|-\|f^-\|$$

and call this expression the **integral** of f over $\mathscr{X}$ with respect to μ. The symbols $d\mu$ will often be omitted when there is no ambiguity. A complex function is integrable if its real and imaginary parts are integrable.

2.2 Properties of the integral

Theorem 6. *If f is integrable, so are f^+, f^-, and $|f|$, and*

$$\int f = \|f^+\| - \|f^-\| = \int f^+ + \int f^-,$$

$$\int |f| = \|f\| = \|f^+\| + \|f^-\| = \int f^+ - \int f^-.$$

The first part follows at once from Theorem 3, with $f_n = \theta_n$, and the definition of the integral. In the second part, we have only to prove that

$$\|f\| = \|f^+\| + \|f^-\|$$

and this follows from Theorem 3 and the identity

$$\|\theta_n\| = \|\theta_n^+\| + \|\theta_n^-\|$$

of Theorem 5.

Theorem 7. *A simple function θ is integrable and*

$$\int \theta = A(\theta).$$

This follows at once from the definition and Theorem 5.

Theorem 8. *If θ_n is simple and $\theta_n \uparrow \lambda$, $\|\lambda\| < \infty$, then $\|\lambda - \theta_n\| \to 0$, λ is integrable and*

$$\int \lambda = \lim \int \theta_n.$$

We note first that $\lim A(\theta_n)$ exists since $A(\theta_n)$ increases and is bounded by $\|\lambda\|$. Then since $\theta_m - \theta_{m-1}$ is non-negative and simple and

$$\lambda - \theta_n = \sum_{m=n+1}^{\infty} (\theta_m - \theta_{m-1}),$$

we deduce from Theorems 2 and 5 that

$$\|\lambda - \theta_n\| \leqslant \sum_{m=n+1}^{\infty} A(\theta_m - \theta_{m-1}) = \sum_{m=n+1}^{\infty} \{A(\theta_m) - A(\theta_{m-1})\}$$

$$= \lim_{M \to \infty} A(\theta_M) - A(\theta_n),$$

and the first part follows. The second part comes from Theorems 3, 5, and 7.

Corollary. *If* $\|f\|<\infty$, *we can define an integrable function* λ *so that* $|f|\leq\lambda$.

Theorem 9. *If* f_n *is integrable and* $\|f-f_n\|\to 0$, *then* f *is integrable and*

$$\int f_n\to\int f,\quad \int f_n^+\to\int f^+,\quad \int f_n^-\to\int f^-,\quad \int|f_n|\to\int|f|.$$

By the definition of the integral, we can define simple functions θ_n so that $\|f_n-\theta_n\|\to 0$ Then

$$\|f-\theta_n\|=\|f-f_n+f_n-\theta_n\|\leq\|f-f_n\|+\|f_n-\theta_n\|=o(1)$$

and f is integrable. The second part follows from Theorem 3 and the definition of the integral.

Theorem 10. *If* $\|f\|=0$, *then* f *is integrable and* $\int f=\int|f|=0$.

We need only take $\theta_n=0$ in the definition. A function of this type is called a **null-function.**

Theorem 11. *If* f *is real and integrable,* $|\int f\,\mathrm{d}\mu|\leq\int|f|\,\mathrm{d}\mu$. *If* f *is complex and integrable, so is* $|f|$ *and the same inequality holds, and the same is true if the values of* f *are real vectors.*

If f is real, $|\int f\,\mathrm{d}\mu|=|\|f^+\|-\|f^-\||\leq\|f^+\|+\|f^-\|=\int|f|\,\mathrm{d}\mu$. If f is complex and integrable, we can define complex valued simple functions θ_n so that $\|f-\theta_n\|\to 0$, and the conclusion follows from the inequalities $\|\,|f|-|\theta_n|\,\|\leq\|\,|f-\theta_n|\,\|$ and $|A(\theta_n)|\leq A(|\theta_n|)$.

Theorem 12. *A finite linear combination* $\sum a_nf_n$ *of integrable functions* f_n *is integrable (provided that it is defined for all* x*) and*

$$\int\sum a_nf_n=\sum a_n\int f_n$$

This follows at once from Theorem 9 and the linearity property (2) of Section 2.1 for the approximating simple function.

Corollary. *If* f *and* g *are integrable, so are* $\sup(f,g)=f+(g-f)^+$ *and* $\inf(f,g)=f+(g-f)^-$.

Theorem 13. *If* f *and* g *are integrable and* $g\leq f$, *then* $\int g\leq\int f$. *In particular,* $\int f\geq 0$ *if* $f\geq 0$.

The special case is obvious from the definition, and the general inequality follows from Theorem 12, since

$$\int f=\int\{g+(f-g)\}=\int g+\int(f-g)$$

and $f-g\geqslant 0$. As an immediate corollary, we have

Theorem 14 (Mean-value Theorem). *If f is integrable and if $c\leqslant f\leqslant C$ in a simple set I and $f=0$ outside I, then*

$$c\mu(I)\leqslant \int f\leqslant C\mu(I).$$

Theorem 15. *If $\{f_n\}$ is a finite set of integrable functions, and $\sum f_n$ is defined for all x, then*

$$\left|\int \sum f_n\right|\leqslant \sum \int |f_n|.$$

For $|\int\sum f_n|\leqslant\int|\sum f_n|\leqslant\int\sum|f_n|=\sum\int|f_n|$ by Theorems 11, 13, and 12.

The theorems above show that the integrable functions form a linear space L which is also a metric space with $\|f\|$ as metric. We shall write $f\varepsilon L$ (or $f\varepsilon L_\mu$ when the specification is necessary) to mean that f is integrable.

We define f_X for a function f and a set X to be the function which is equal to f in X and vanishes outside. If γ is the unit function taking the value 1 at every point of $\mathscr{X}$, γ_X is called the **characteristic function** of the set X and takes the values 1 or 0 according as x is in X or X'. A set X is called a **null set** or a **set of zero measure** if γ_X is a null function, so that

$$\|\gamma_X\|=\int \gamma_X=0.$$

Any subset of a null set is obviously a null set. A property which holds outside a null set is said to hold **almost everywhere** (abbreviated to a.e.) In particular, we shall write $f\geqslant g$ a.e. if $f(x)\geqslant g(x)$ for almost all x, and $f_n\to f$ a.e. if $f_n(x)\to f(g)$ for almost all x in X.

Theorem 16. *The sum of a sequence of null functions is a null function provided that the sum is defined for all x. The union of a sequence of null sets is a null set.*

The first part follows immediately from Theorem 2, and it is sufficient to prove the second part in the case when the null sets X_n are disjoint and have union X. Then $\gamma_X=\sum\gamma_{X_n}$, and the conclusion follows from the first part.

Theorem 17. (i) *If f is integrable, $|f|<\infty$ a.e.* (ii) *If f is null, $f=0$ a.e.* (iii) *Conversely, if $f=0$ a.e., then f is null.* (iv) *An integrable function remains integrable and has the same integral after its values (including $\pm\infty$) are changed in a null set.*

If $c>0$, let γ_c be the characteristic function of the set in which $|f|\geqslant c$, so that γ_∞ denotes the characteristic function of the set in which $f(x)=$

$\pm\infty$. Then since $\gamma_\infty \leqslant \gamma_c \leqslant |f|/c$ for $0 < c < \infty$, we have

$$\|\gamma_\infty\| \leqslant \|\gamma_c\| \leqslant c^{-1}\|f\|,$$

and, on letting $c \to \infty$, we get $\|\gamma_\infty\| = 0$, which gives (i). But if f is null, $\|f\| = 0$, and the last inequality gives $\|\gamma_c\| = 0$ for $c > 0$. The set in which $f \neq 0$ is the union of the sets γ_c with $c = 1, \frac{1}{2}, \frac{1}{3}, \ldots$, and is therefore null by Theorem 16.

It is sufficient to prove (iii) in the case when $|f| = \infty$ in a null set X and $f = 0$ in X'. Then if we define functions $\lambda_m = \gamma_X$ for $m = 1, 2, 3, \ldots$, we have $\|\lambda_m\| = 0$, and since $\lambda_m = \gamma_X = 1$ in X, it follows that $|f| = \infty = \lambda = \sum \lambda_m$ in X, while $|f| = 0 \leqslant \lambda$ in X'. Hence,

$$|f| \leqslant \lambda, \qquad \|f\| \leqslant \|\lambda\| \leqslant \sum \|\lambda_m\| = 0.$$

For (iv), suppose that $f = g$ outside a null set and let $f_1 = f$, $g_1 = g$ when $f = g$ and f and g are both finite, and let $f_1 = g_1 = 0$ in the remaining null set. Then $f_1 = g_1$, $f = f_1 + (f - f_1)$, $g = g_1 + (g - g_1)$ for all x, and it follows from Theorem 12 and (i) and (iii) above that $\int f = \int g$.

An importance consequence of this theorem is that we can ignore the values of a function or sequence of functions in any null set. In fact, the functions need not be defined in a null set, and the preceding theorems, particularly Theorems 12, 13, 14, and 15, can be generalized in this sense.

Theorem 18 (Convergence Theorem). *Suppose that $f_n \in L$ for $n = 1, 2, \ldots$, $|f_n| \leqslant \lambda$, $\lambda \in L$, and that $f_n \to f$* a.e. *Then $f \in L$ and*

$$\int f_n \to \int f, \qquad \int |f_n - f| \to 0.$$

Moreover, the theorem remains valid if n is replaced by a continuous real variable t in an interval $-\infty \leqslant a < t < b \leqslant \infty$ and the limit is taken as $t \to a$ or $t \to b$.

We prove as lemmas two special cases of the theorem.

Lemma 1. *If ξ_n is simple, $\xi_n \geqslant \xi_{n+1} \geqslant 0$ and $\xi_n \to 0$* a.e., *then $\int \xi_n \to 0$.*

We can write $\xi_1 - \xi_n \uparrow \beta$, where β is defined for all x since ξ_n is monotonic. Since $\xi_1 - \xi_n$ is simple and β, the sum of ξ_1 and null function, is integrable, it follows from Theorems 8 and 12 that

$$\int \xi_1 - \int \xi_n = \int (\xi_1 - \xi_n) \to \int \beta = \int \xi_1,$$

and the conclusion follows.

Lemma 2. *Suppose that, for each n, λ_n is the limit of an increasing* *sequence of non-negative simple functions, that* $\|\lambda_1\|<\infty$, λ_n *decreases for all* x *and* $\lambda_n \to 0$ *a.e. Then* $\|\lambda_n\| \to 0$.

Let $\varepsilon>0$, $\varepsilon_m>0$, $\sum \varepsilon_m=\varepsilon$. Then after Theorem 8, we can define simple functions η_m so that $\|\lambda_m-\eta_m\|<\varepsilon_m$, $0\leqslant\eta_m\leqslant\lambda_m$. Then

$$\xi_n = \inf_{m\leqslant n} \eta_m$$

is simple, ξ_n decreases and since $0\leqslant\xi_n\leqslant\eta_n\leqslant\lambda_n$, it follows that $\xi_n \downarrow 0$ a.e. and therefore, by Lemma 1, $\|\xi_n\|=\int \xi_n \to 0$. But λ_m decreases and

$$\lambda_n = \inf_{m\leqslant n} \lambda_m = \inf_{m\leqslant n} \{\eta_m+(\lambda_m-\eta_m)\}\leqslant\xi_n+\sum_{m=1}^{n}(\lambda_m-\eta_m),$$
$$\|\lambda_n\|\leqslant\|\xi_n\|+\sum_{m=1}^{n}\|\lambda_m-\eta_m\|\leqslant\varepsilon+o(1), \tag{1}$$

and the conclusion follows.

Proof of Theorem 18. *If* $\varepsilon>0$, $\varepsilon_v>0$, $\sum \varepsilon_v=\varepsilon$, we define simple functions θ_v so that

$$\|f_v-\theta_v\|\leqslant\varepsilon_v. \tag{2}$$

Then

$$\sup_{v\geqslant n}|f_v-\theta_v|\leqslant\sum_{v=n}^{\infty}|f_v-\theta_v|$$

$$\|\sup_{v\geqslant n}|f_v-\theta_v|\|\leqslant\sum_{v=n}^{\infty}\|f_v-\theta_v\|\leqslant\sum_{v=n}^{\infty}\varepsilon_v=o(1)$$

as $n\to\infty$. But

$$\limsup|f_v-\theta_v|\leqslant\sup_{v\geqslant n}|f_v-\theta_v|$$

for all n, and it follows that $\limsup|f_v-\theta_v|$ is a null function and vanishes a.e. Since $f_v\to f$ a.e., it then follows that $\theta_v\to f$ a.e. and, by the general principle of convergence, that

$$\lambda_n=\sup_{i,j\geqslant n}|\theta_i-\theta_j|\to 0 \quad \text{a.e.}$$

Moreover, λ_n obviously decreases for every x:

$$\lambda_1\leqslant 2\sup|\theta_v|\leqslant 2\sup|f_v|+2\sup|f_v-\theta_v|\leqslant 2\lambda+2\sum|f_v-\theta_v|,$$
$$\|\lambda_1\|\leqslant 2\,\|\lambda\|+\varepsilon<\infty,$$

and λ_n therefore satisfies the conditions of Lemma 2, and $\|\lambda_n\|\to 0$. But since

$$|f-\theta_n| = \lim_{i\to\infty} |\theta_i - \theta_n| \leq \lambda_n \quad \text{a.e.},$$

it follows that $\|f-\theta_n\| \to 0$. This shows that $f \in L$ and the conclusion follows from this and (2) and Theorems 2 and 3.

The analogous form for the continuous variable t follows immediately since $\int f_t \to \int f$ and $\int |f_t - f| \to 0$ as $t \to b$ through any sequence of values, and this is true only if the limits exist as $t \to b$ through all values.

The following theorem for series follows as an immediate corollary of Theorem 18.

Theorem 19. *Suppose that* $a_n \in L$ *for* $n = 1, 2, \ldots,$

$$\left| \sum_{v=1}^{n} a_v \right| \leq \lambda,$$

$\lambda \in L$, *and that*

$$\sum_{v=1}^{\infty} a_v = s \quad \text{a.e.}$$

Then s *is integrable and*

$$\int s = \sum_{v=1}^{\infty} \int a_v.$$

The following convergence theorem, though weaker than Theorem 18, is often useful.

Theorem 20. *Suppose that* $f_n \in L$ *for* $n = 1, 2, \ldots,$ *and* $f_n \uparrow f$ *a.e. Then*

$$\lim_{n\to\infty} \int f_n = \int f$$

in the sense that if one side exists, so does the other, and the two are equal. Moreover, the theorem remains valid if n *is replaced by a continuous variable as in Theorem* 18.

The existence of the right-hand side means that $f \in L$ and therefore $|f_n| \leq \sup\{|f_1|, |f|\}$ and the conclusion follows from Theorem 18.

Conversely, if the left-hand side is finite,

$$\begin{aligned} \|f\| &= \|f_1 + \sum_{v=2}^{\infty} (f_v - f_{v-1})\| \leq \|f_1\| + \sum_{v=2}^{\infty} \|f_v - f_{v-1}\| \\ &= \|f_1\| + \lim_{n\to\infty} \sum_{v=2}^{n} \left(\int f_v - \int f_{v-1} \right) \\ &= \|f_1\| - \int f_1 + \lim_{n\to\infty} \int f_n < \infty, \end{aligned}$$

by Theorems 2 and 12, and the conclusion follows from Theorem 18 and the corollary of Theorem 8.

2.3 Measurability and measure

A **function** is **measurable** if it is the limit a.e. of a sequence of simple functions.

A simple function is obviously measurable and so is the unit function γ, since it is the limit of characteristic functions of a sequence of simple sets whose limit is $\mathscr{X}$. The following theorems show how measurability is preserved under the familiar analytic operations.

Theorem 21. *The modulus and positive and negative components of a measurable function are measurable. If two functions f and g are measurable, so are $f \pm g$, fg, $\sup(f, g)$, $\inf(f, g)$, and f/g, provided that they are defined a.e.*

All these results except the last follow immediately from the fact that the operations applied to simple functions produce simple functions. In the last part, we can suppose that $f = 1$ and define simple functions θ_n so that $\theta_n \to g$ a.e. We then define simple sets I_n so that $I_n \to \mathscr{X}$ and define the simple functions ξ_n by $\xi_n = \theta_n^{-1}$ if $\theta_n \neq 0$, $\xi_n = 1$ when $\theta_n = 0$, $x \in I_n$, and $\xi_n = 0$ when $\theta_n = 0$, $x \in I'_n$. It is plain that ξ_n is simple and $\xi_n \to g^{-1}$ a.e.

Theorem 22. *In order that f be integrable it is necessary and sufficient that it be measurable and $\|f\| < \infty$.*

Necessity is clear from the first part of the proof of Theorem 18 (with $f_v = f$). Conversely, we suppose that θ_n is simple, $\theta_n \to f$ a.e., and (using the corollary of Theorem 8) that $|f| \leq \lambda$, $\lambda \in L$. If we let

$$\lambda_n = \theta_n \text{ if } |\theta_n| \leq \lambda, \qquad \lambda_n = \lambda \text{ if } \theta_n > \lambda, \qquad \lambda_n = -\lambda \text{ if } \theta_n < -\lambda,$$

it follows that $|\lambda_n| \leq \lambda$ and λ_n is integrable by the corollary of Theorem 12. Since $\lambda_n \to f$ a.e., the conclusion follows from Theorem 18. It is often convenient, when measurability can be assumed, to use the notation $\|f\| < \infty$ as synonymous with $f \in L$.

Theorem 23. *If f_n is measurable for $n = 1, 2, 3, \ldots$, so are $\limsup f_n$ and $\liminf f_n$. In particular, if $f_n \to f$ a.e., then f is measurable.*

We prove the theorem first in the special case $f_n \to f$ a.e. We can subdivide $\mathscr{X}$ into a sequence of disjoint simple sets I_j with

$$0 < \mu(I_j) < \infty$$

and the function h defined to be equal to

$$\{j^2 \mu(I_j)\}^{-1}$$

in I_j is positive and integrable by Theorem 20.

Since f_n is measurable, we can define simple functions θ_{nv} a.e., each vanishing outside a finite union of sets I_j, so that

$$\lim_{v\to\infty} \theta_{nv} = f_n \quad \text{a.e.}$$

for each n. Then

$$g_{nv} = \frac{h\theta_{nv}}{h+|\theta_{nv}|} \to g_n \quad \text{a.e.,}$$

where

$$g_n = \frac{hf_n}{h+|f_n|} \quad \text{if} \quad f_n \neq \pm\infty, \qquad g_n = \pm h \quad \text{if} \quad f_n = \pm\infty,$$

and g_n is measurable since g_{nv} is simple. Also $g_n \to g$ a.e. where

$$g = \frac{hf}{h+|f|} \quad \text{if} \quad f \neq \pm\infty, \qquad g = \pm h \quad \text{if} \quad f = \pm\infty.$$

But $|g_n| \leq h$, and h is integrable, and it therefore follows from Theorems 18 and 22 that g is measurable, and we can define simple functions θ_n, each vanishing outside a finite union of sets I_j, so that $|\theta_n| < h$, $\theta_n \to g$ a.e. Then $\theta_n h(h-|\theta_n|)^{-1}$ is simple and tends almost everywhere to f, and it follows that f is also measurable.

In the general case, it follows from Theorem 21 and what we have just proved that

$$\sup_{v\geq n} f_v = \lim_{m\to\infty} \sup_{m\geq v\geq n} f_v$$

is measurable, and so also

$$\limsup f_n = \lim_{n\to\infty} \sup_{v\geq n} f_v.$$

We can treat $\liminf f_n$ similarly.

A **set** X is **measurable** if its characteristic function γ_X is measurable.

Any simple set is measurable and so is $\mathscr{X}$. If X is measurable and $\|\gamma_X\| < \infty$, then γ_X is integrable by Theorem 22 and we define the **measure** $\mu(X)$ of X by $\mu(X) = \int \gamma_X$. If X is measurable and $\|\gamma_X\| = \infty$, we say that X has infinite measure and write $\mu(X) = \infty$.

The notation $\mu(X)$ is consistent with that of the original simple measure $\mu(I)$ on the simple sets, so that we may think of $\mu(X)$ as an extension of the simple measure to a wider system of sets. In particular, a null set is simply a measurable set of zero measure.

If $f \in L$ and X is measurable, $f\gamma_X$ is integrable by Theorems 21 and 22 and its integral is called the **integral of** f **over** X and written

$$F(X) = \int_X f.$$

The essential properties of measurable sets and integrals over them can be summarized as follows.

Theorem 24. (i) *The measurable sets form a σ-ring on which $\mu(X)$ is a measure and an extension of the simple measure $\mu(I)$; and $\mu(\mathscr{X})<\infty$ if and only if $\mu(I)$ is bounded on simple sets.* (ii) *If $f\in L$, then*

$$F(X)=\int_X f,\qquad F^+(X)=\int_X f^+,\qquad F^-(X)=\int_X f^-$$

are completely additive on the σ-ring of measurable sets. (iii) *$F^+(X)$, $F^-(X)$ are the Hahn–Jordan components of $F(X)$ in the sense of Chapter* 1, *and if $F(X)=0$ for all measurable X, then f is null.*

If $f\geqslant 0$ is measurable and X, X_1, X_2 are measurable sets, $f_X=f\gamma_X$ is measurable by Theorem 21 and

$$f_{X'}=f-f_X,\qquad f_{X_1\cup X_2}=\sup[f_{X_1},f_{X_2}].$$

If $X_1, X_2, \ldots$ are measurable and disjoint, and $X=\bigcup X_\upsilon$, then $f_X=\sum f_{X_\upsilon}$. In particular, we may substitute $f=\gamma$ and the first part of (i) follows at once from the definitions and Theorems 21, 22, and 23. The second part comes from the fact that $I_n\to\mathscr{X}$ and so $\mu(I_n)\to\mu(\mathscr{X})$ for some sequence of simple sets I_n. If f is integrable, (ii) follows for f^+ from Theorems 12 and 18, and the conclusions extend at once to f^- and f.

For (iii) we need only observe that $F^+(X)$ and $F^-(X)$ have the properties of the Hahn–Jordan components and that this decomposition is unique by Theorem 3 of Chapter 1. Finally, if $F(X)=0$ for all measurable X, it follows that $F^+(\mathscr{X})=F^-(\mathscr{X})=0$ and therefore f^+, f^-, and f are all null.

The Borel extension of the simple sets depends only on the sets and not on any measure defined in them. Since it is the minimal extension, it is contained in the σ-ring of measurable sets defined by any simple measure $\mu(I)$, and every Borel set is measurable. The converse is not generally true, although the following theorems show that a measurable set is almost a Borel set in a certain sense. We say that f is a **Borel function** if† $\{x:f(x)>c\}$ is a Borel set for every real c.

Theorem 25. (i) *In order that f be measurable it is necessary and sufficient that $\{x:f(x)>c\}$ be measurable for every real c.* (ii) *A Borel function is measurable and every measurable function is equal a.e. to a Borel function.* (iii) *A measurable set differs from a Borel set by a null set, and any null set is contained in a Borel null set.*

† In the usual notation, $\{x:P(x)\}$ is the set of points x for which the property $P(x)$ holds.

First suppose that $\{x: f(x) > c\}$ is measurable for every c. Then

$$X_\infty = \{x: f(x) = \infty\} = \bigcap_{v=1}^{\infty} \{x: f(x) > v\} \quad \text{and} \quad X_{-\infty} = \{x: f(x) = -\infty\}$$

are measurable and

$$X_{nv} = \{x: v+1 \geqslant nf(x) > v\} = \{x: nf(x) > v\} \cap \{nf(x) \leqslant v+1\}$$

is measurable for $n = 1, 2, 3, \ldots,$ $v = 0, \pm 1, \pm 2, \ldots,$ and so are the characteristic functions γ_∞, $\gamma_{-\infty}$, γ_{nv} of X_∞, $X_{-\infty}$, X_{nv}. The function

$$g_n = \frac{1}{n} \sum_{v=-\infty}^{\infty} v\gamma_{nv} + n\gamma_\infty - n\gamma_{-\infty}$$

is also measurable and $g_n \to f$ a.e., and this proves the sufficiency part of (i) and the first part of (ii).

Conversely, if f is measurable, we define simple functions θ_n so that $\theta_n \to f$ a.e. Then if $f_1 = \liminf \theta_n$,

$$\{x: f_1(x) > c\} = \bigcup_v \{x: \theta_n(x) \geqslant c + 1/v \text{ except for a finite set of } n\}$$

$$= \bigcup_v \liminf_{n\to\infty} \{x: \theta_n(x) \geqslant c + 1/v\}$$

is obtained by countable set operations on the simple sets

$$\{x: \theta_n(x) \geqslant c + 1/v\}$$

and is therefore a Borel set. Hence f_1 is a Borel function and $f - f_1$ is null since $\theta_n \to f$ a.e. This establishes the second part of (ii), and (iii) follows from this and the fact that $X = \{x: \gamma_X(x) > \frac{1}{2}\}$. Finally, any null set is contained in a decreasing sequence of simple sets with measures tending to 0, and is contained in their intersection, which is a Borel null set.

Theorem 26. *The measurable sets derived from a simple measure $\mu(I)$ form the minimal σ-ring which contains the simple sets and all null sets; and $\mu(X)$ is the only possible extension of $\mu(I)$ to this σ-ring. We shall call $\mu(X)$ the Lebesgue extension of $\mu(I)$.*

Corollary. *If the simple measure is itself defined over a σ-ring, its Lebesgue extension is the measure obtained by adding the null sets (if any) which are not already included as simple sets. Moreover, the integral derived from a simple measure μ is identical with that obtained by replacing μ by its Lebesgue extension and treating the latter as a simple measure.*

The first part follows at once from Theorem 25(iii). For the second part, we note that in the proof of Theorem 25 with $f = \gamma_X$, the measure of $\{x: \theta_n(x) \geqslant \frac{1}{2} + 1/v\}$ is defined uniquely by the simple measure and the same is true, by the continuity of μ, for $\{x : f_1(x) > \frac{1}{2}\}$. This differs by a null set from X and it is therefore sufficient to show that if μ^* is any other measure extension of $\mu(I)$, then $\mu^*(X) = 0$ when $\mu(X) = 0$. This follows from the fact that if $\mu(X) = 0$, $\varepsilon > 0$, we can define simple sets I_j so that $X \subset \bigcup I_j$, $\sum \mu(I_j) < \varepsilon$, and then $\mu^*(\bigcup I_j) \leqslant \sum \mu^*(I_j) = \sum \mu(I_j) < \varepsilon$.

2.4 Absolutely continuous set functions and their derivatives

A completely additive set function F defined on a σ-ring is called **absolutely continuous** with respect to a measure μ on the same σ-ring if $F(X) \to 0$ uniformly as $\mu(X) \to 0$. In other words, if $\varepsilon > 0$, we can define $\delta(\varepsilon) > 0$ so that $|F(X)| < \varepsilon$ for all measurable sets X with $\mu(X) < \delta$.

In fact, when μ is derived from a simple measure, it is enough to have this condition for simple sets. For if X is measurable and we suppose that $F(X)$ is non-negative and $\mu(X) \leqslant \delta/4$, we can use Theorem 8 and the definition of L to define simple functions θ_n so that $\theta_n \uparrow \lambda$, $\gamma_X \leqslant \lambda$, $\int \lambda \, d\mu \leqslant \delta/2$; and it follows that the simple sets $I_n = \{x : \theta_n(x) > \frac{1}{2}\}$ increase to a limit J which contains X. By the definition of I_n, $\mu(I_n)/2 \leqslant \int \theta_n \, d\mu \leqslant \int \lambda \, d\mu \leqslant \delta/2$, and then $F(I_n) \leqslant \varepsilon$ and $F(X) \leqslant F(J) = \lim F(I_n) \leqslant \varepsilon$ by the complete additivity of F.

Theorem 27. *A completely additive set function F defined on a σ-ring is absolutely continuous if and only if $F(X) = 0$ for all null sets X. In particular, if $f \in L$ the function F defined by $F(X) = \int_X f$ is absolutely continuous and $F(X) = 0$ for every null set X.*

Suppose first that F is absolutely continuous and $\mu(X) = 0$, so that X is contained in a decreasing sequence of sets X_n for which $\mu(X_n) \to 0$. Then $F(X) \leqslant \lim F(X_n) = 0$. Conversely, if F is not absolutely continuous, there is a sequence of sets X_n and $\delta > 0$ such that $F(X_n) \geqslant \delta$, $\sum \mu(X_n) < \infty$. Then if $X = \limsup X_n$, we have $F(X) \geqslant \delta$, while $\mu(X) \leqslant \lim\limits_{m\to\infty} \sum\limits_{n \geqslant m} \mu(X_n) = 0$.

If $f \in L$ and $\varepsilon > 0$, we define a simple function θ so that $\|f - \theta\| < \varepsilon/2$. Then if $C = \sup|\theta|$ and X is measurable, $|F(X)| \leqslant \int_X |f| \leqslant \int_X |\theta| + \|f - \theta\| \leqslant C\mu(X) + \varepsilon/2 \leqslant \varepsilon$ for $\mu(X) \leqslant \varepsilon/2C$, and F is absolutely continuous.

We have proved in Theorems 24 and 27 that the integral $F(X)$ of an integrable function is absolutely continuous and completely additive. The next theorem gives a converse to this and also gives some information about the structure of more general set functions which are not necessarily absolutely continuous.

Theorem 28 (Radon–Nikodym Theorem). *Suppose that H is completely additive in the σ-ring of sets Borel measurable with respect to a measure μ in $\mathscr{X}$. Then there is a unique decomposition $H(X) = F(X) + Q(X)$ in which* *the **singular component** $Q(X)$ of $H(X)$ satisfies $Q(X)=Q(X\cap S)=H(X\cap S)$ for some fixed null Borel set S, $F(X)$ is absolutely continuous and*

$$F(X)=\int_X f\,\mathrm{d}\mu = H(X-S)$$

for some function f (unique except for a null function) which is integrable over every X for which $H(X)$ is finite and is called the Radon–Nikodym derivative of F.

In particular, if H itself is absolutely continuous, $Q(X)=0$ and $H(X)=F(X)=\int_X f\,\mathrm{d}\mu$. Moreover, if $H(X)\geqslant 0$, then $F(X)\geqslant 0$ and $f(x)\geqslant 0$ a.e.

Since we have assumed in our definition of simple sets that $\mathscr{X}$ is a countable union of simple sets of finite measure, it is enough to prove the theorem when $\mu(\mathscr{X})<\infty$, and after the Hahn–Jordan theorem, we may suppose that $H(X)\geqslant 0$.

We define $\boldsymbol{T}$ to be the class of integrable functions $t\geqslant 0$ for which

$$T(X)=\int_X t\,\mathrm{d}\mu \leqslant H(X)$$

for all B-measurable X. We then define a sequence t_n in $\boldsymbol{T}$ so that

$$T_n(\mathscr{X})\to T^*=\sup_{t\in\mathbf{T}} T(\mathscr{X})\leqslant H(\mathscr{X})<\infty. \tag{1}$$

If we now define

$$p_n=\sup_{\upsilon\leqslant n} t_\upsilon, \qquad X_\upsilon=\{x: t_\upsilon=p_n,\ t_j<p_n \text{ for } j=1,2,\ldots,\upsilon-1\},$$

it follows that X_υ are disjoint and measurable (by Theorem 25) and $X=\bigcup X_\upsilon$,

$$P_n(X)=\int_X p_n\,\mathrm{d}\mu=\sum_{\upsilon=1}^{n} P_n(X_\upsilon)\leqslant \sum_{\upsilon=1}^{n} H(X_\upsilon)=H(X),$$

and therefore $p_n\in\boldsymbol{T}$. But p_n increases and has a limit f which is also in $\boldsymbol{T}$, by Theorem 20, and therefore $F(X)\leqslant H(X)$, $Q(X)=H(X)-F(X)\geqslant 0$ for every measurable X. Also, since $T_n(\mathscr{X})\leqslant P_n(\mathscr{X})\leqslant T^*$, it follows from (1) that $P_n(\mathscr{X})\to T^*$ and from Theorem 20 that $P_n(\mathscr{X})\to F(\mathscr{X})$ and so

$$F(\mathscr{X})=T^*. \tag{2}$$

Now let

$$Q_n(X)=Q(X)-\mu(X)/n,$$

and let $\mathscr{X}_n^+$ and $\mathscr{X}_n^-$ be the B-sets defined for $Q_n(X)$ by the Hahn–Jordan theorem, so that $Q(X)\geqslant\mu(X)/n$ for $X\subset\mathscr{X}_n^+$, $Q(X)\leqslant\mu(X)/n$ for $X\subset\mathscr{X}_n^-$.

Then

$$H(X) \geqslant F(X) + \frac{\mu(X)}{n} = \int_X \left(f + \frac{1}{n}\right) \mathrm{d}\mu \quad \text{for} \quad X \subset \mathscr{X}_n^+$$

and this shows that the function equal to f in $\mathscr{X}_n^-$ and to $f + 1/n$ in $\mathscr{X}_n^+$ belongs to $\boldsymbol{T}$ and therefore, after (2),

$$\frac{1}{n}\mu(\mathscr{X}_n^+) + F(\mathscr{X}) = \int_{\mathscr{X}_n^+} \left(f + \frac{1}{n}\right) \mathrm{d}\mu + \int_{\mathscr{X}_n^-} f \,\mathrm{d}\mu \leqslant T^* = F(\mathscr{X}).$$

This implies that $\mu(\mathscr{X}_n^+) = 0$ for each n, and $S = \bigcup_n \mathscr{X}_n^+$ is null. Since $X - S \subset \mathscr{X}_n^-$ for every n, it follows that $Q(X - S) = 0$ and $Q(X) = Q(X \cap S) + Q(X - S) = Q(X \cap S) = H(X \cap S) - F(X \cap S) = H(X \cap S)$ since $X \cap S$ is null. Likewise, $F(X) = F(X - S) + F(X \cap S) = F(X - S) = H(X - S) - Q(X - S) = H(X - S)$.

To prove uniqueness of $F(X)$ and $Q(X)$, we suppose that the decomposition can be done in two ways and

$$H(X) = F_1(X) + Q_1(X) = F_2(X) + Q_2(X),$$

where $F_1(X)$, $F_2(X)$ are integrals and $Q_1(X)$, $Q_2(X)$ vanish on all sets disjoint from the null sets S_1, S_2 whose union S is also null. Then

$$F_1(X) = F_1(X - X \cap S), \qquad F_2(X) = F_2(X - X \cap S),$$
$$F_1(X) - F_2(X) = Q_2(X - X \cap S) - Q_1(X - X \cap S) = 0$$

and the uniqueness of f, and therefore of F and Q, follows from Theorem 24(iii).

Finally, if H is absolutely continuous, so is $Q = H - F$ by Theorem 27, and since an absolutely continuous function vanishes on a null set, $Q(X) = Q(X \cap S) = 0$.

If G is a completely additive set function over a σ-ring of sets X and if $\mathscr{X}$ is the countable union of sets X_v for which $G(X_v)$ is finite, we can decompose G into the two measures G^+ and $-G^-$ defined by the Hahn–Jordan Theorem of Chapter 1 and establish an integral with respect of each of them, using the whole σ-ring on which $G(X)$ is defined as the ring of simple sets. We then say that a function f is **integrable with respect to** G if it is integrable with respect to G^+ and $-G^-$, and we define the integrals $\int f \,\mathrm{d}G$, $\int f\,|\mathrm{d}G|$ by

$$\int f \,\mathrm{d}G = \int f \,\mathrm{d}G^+ - \int f \,\mathrm{d}(-G^-), \qquad \int f\,|\mathrm{d}G| = \int f \,\mathrm{d}G^+ + \int f \,\mathrm{d}(-G^-).$$

The σ-ring of measurable sets will be the original σ-ring with the addition of null sets if they are not already included. Since the two integrals on the right can be treated separately, there is no need for anything more than the theory already established and all the theorems proved for μ-measures can be applied immediately. The following extension of Theorem 11 is perhaps worth stating separately.

Theorem 29. *If f is real or complex an integrable with respect to G, then*

$$\left|\int f\,\mathrm{d}G\right| \leqslant \int |f|\,|\mathrm{d}G|.$$

For,

$$\begin{aligned}\left|\int f\,\mathrm{d}G\right| &= \left|\int f\,\mathrm{d}G^{+} - \int f\,\mathrm{d}(-G^{-})\right| \\ &\leqslant \left|\int f\,\mathrm{d}G^{+}\right| + \left|\int f\,\mathrm{d}(-G^{-})\right| \qquad \text{(by Theorem 11)} \\ &\leqslant \int |f|\,\mathrm{d}G^{+} + \int |f|\,\mathrm{d}(-G^{-}) = \int |f|\,|\mathrm{d}G|.\end{aligned}$$

The following theorem shows that integrals with respect to G may be reduced to integrals with respect to another measure μ when G is absolutely continuous with respect to μ.

Theorem 30. *Suppose that $g \in L$, $G(X) = \int_X g\,\mathrm{d}\mu$ and that f is measurable. Then $\int f\,\mathrm{d}G = \int fg\,\mathrm{d}\mu$ in the sense that if one integral exists, so does the other, and the two are equal.*

It is sufficient, in view of the decomposition $f = f^{+} + f^{-}$, $G = G^{+} + G^{-}$, where $G^{+}(X) = \int_X g^{+}\,\mathrm{d}\mu$, $G^{-}(X) = \int_X g^{-}\,\mathrm{d}\mu$, by Theorem 24(ii), to prove the result when $f \geqslant 0$, $g \geqslant 0$ and define simple sets I_n so that $I_n \uparrow \mathscr{X}$ and functions f_n by

$$f_n = \inf[f, n] \text{ in } I_n, \qquad f_n = 0 \text{ in } I_n'.$$

Then, since f is measurable, we can define a sequence of simple functions θ_{nv} so that

$$0 \leqslant \theta_{nv} \leqslant n \text{ in } I_n, \qquad \theta_{nv} = 0 \text{ in } I_n', \qquad \theta_{nv} \to f_n \quad \text{a.e.}$$

It follows then from Theorem 18 and the fact that the present theorem is obviously true when f is replaced by a simple function that

$$\int f_n\,\mathrm{d}G = \lim_{v\to\infty} \int \theta_{nv}\,\mathrm{d}G = \lim_{v\to\infty} \int \theta_{nv} g\,\mathrm{d}\mu = \int f_n g\,\mathrm{d}\mu.$$

Since $f_n \uparrow f$, the conclusion follows from Theorem 20.

We end this section by showing how a mapping of a measure space $\mathscr{T}$ onto another space $\mathscr{X}$ can be used to induce a measure and integral on the latter. We denote the mapping by $x = \alpha(t)$ and suppose that it maps the whole of $\mathscr{T}$ onto the whole of $\mathscr{X}$ without assuming that it is necessarily one to one.

If X is any set in $\mathscr{X}$, its inverse image $\alpha^{-1}(X)$ is the set of all points in $\mathscr{T}$ for which $\alpha(t) \in X$. Then it is obvious that $X_1, X_2, \ldots,$ are disjoint if and only if their inverse images are disjoint, and that

$$\alpha^{-1}\{\bigcup_v X_v\} = \bigcup_v \alpha^{-1}(X_v), \qquad \alpha^{-1}\{\bigcap_v X_v\} = \bigcap_v \alpha^{-1}(X_v)$$

for finite or countable systems of sets X_v, whether disjoint or not. It follows that the sets X whose inverse images belong to a σ-ring in $\mathscr{T}$ form a σ-ring in $\mathscr{X}$ and that the one-to-one corresponding between X and $\alpha^{-1}(X)$ determines an isomorphism between the two σ-rings with respect to set operations. In particular, if we have a measure ν in $\mathscr{T}$, the sets X whose inverse images in $\mathscr{T}$ are ν-measurable form a σ-ring in $\mathscr{X}$, and if we define

$$\mu(X) = \nu\{\alpha^{-1}(X)\}$$

whenever $\alpha^{-1}(X)$ is μ-measurable, it is easy to verify that $\mu(X)$ is a measure in $\mathscr{X}$. We call it the **measure induced** in $\mathscr{X}$ by ν and α. We can then proceed to derive the integral in $\mathscr{X}$ induced by ν and α by treating the induced measurable sets in $\mathscr{X}$ as simple sets, and the relationship between integration in $\mathscr{T}$ and $\mathscr{X}$ is shown in the following theorem.

Theorem 31. *Suppose that ν is a measure in $\mathscr{T}$ and μ the measure induced in $\mathscr{X}$ by a mapping $x = \alpha(t)$ of $\mathscr{T}$ onto $\mathscr{X}$. Then*

$$\int f(x)\,\mathrm{d}\mu = \int f[\alpha(t)]\,\mathrm{d}\nu,$$

in the sense that if one integral exists, so does the other, and the two are equal.

After the Corollary of Theorem 26, the measurable sets in $\mathscr{T}$ may be taken as the simple sets through which the integral in $\mathscr{T}$ is defined.

The integral in $\mathscr{X}$ is obtained by defining a simple function θ in $\mathscr{X}$ as one which takes constant values a_j in disjoint sets X_j whose inverse images $T_j = \alpha^{-1}(X_j)$ are measurable in $\mathscr{T}$, and θ then defines uniquely the simple function with values $\theta[\alpha(t)]$ in $\mathscr{T}$. If A_ν, $\|\ \|_\nu$, A_μ, $\|\ \|_\mu$ are defined in the two spaces as in Section 2.1, it follows that

$$A_\mu(\theta) = \sum a_j\mu(X_j) = \sum a_j\nu[\alpha^{-1}(X_j)] = \sum a_j\nu(T_j) = A_\nu[\theta[\alpha(T)]],$$
$$\|f\|_\mu = \|f[\alpha(t)]\|_\nu.$$

If $\int f(x)\,\mathrm{d}\mu$ exists, and we express the condition for this in terms of simple functions and consider the associated simple function in $\mathscr{T}$, it follows at once that $\int f[\alpha(t)]\,\mathrm{d}\nu$ exists and has the same value.

In proving the converse, we have to remember that the measurable sets, or even the simple sets, in $\mathscr{T}$ may not all be inverse images of sets in $\mathscr{X}$ and it is therefore not enough to consider approximation to $\int f[\alpha(t)]\,\mathrm{d}\nu$ by the most general kind of simple function in $\mathscr{T}$. In fact, this difficulty can be overcome by noting that

$$\{t : f[\alpha(t)] > c\} = \alpha^{-1}\{x : f(x) > c\}$$

for any real c, and the function with values $f[\alpha(t)]$ is therefore measurable and integrable with respect to the measure ν on the restricted σ-ring of measurable sets in $\mathscr{T}$ which are also inverse images of sets in $\mathscr{X}$. The correspondence between simple functions can be used again and the conclusion follows.

It is important to notice that the measure induced in $\mathscr{X}$ by the procedure described here may bear no relation to intrinsic properties of $\mathscr{X}$ and may, in fact, not be a useful measure. What is needed in practice is an induced measure which includes among its measurable sets all those sets in $\mathscr{X}$ which arise naturally. For example, if $\mathscr{X}$ is $\mathscr{R}$, it would be most desirable that all Borel sets should be measurable and this can be ensured quite simply by demanding that the function $x = \alpha(t)$ should be measurable with respect to the measure ν in $\mathscr{T}$.

As an example of the theory, suppose that $\mathscr{T} = \mathscr{X} = \mathscr{R}^k$, that the mapping of $\mathscr{T}$ on $\mathscr{X}$ is defined by the non-singular matrix transformation $x = Ct$, and that ν in $\mathscr{T}$ is the classical Lebesgue extension of the volume of figures which is treated in detail in Chapter 3. Then every figure in $\mathscr{X}$ has an inverse image in $\mathscr{T}$ which is a parallelepiped and therefore ν-measurable, and μ is therefore defined for every Lebesgue measurable set in $\mathscr{X}$. Moreover, it is easy to see by elementary coordinate geometry that μ is absolutely continuous with respect to the Lebesgue measure m in $\mathscr{X}$ and has constant Radon–Nikodym derivative equal to $|C|^{-1}$, the ratio between the volumes of corresponding parallelepipeds in $\mathscr{T}$ and $\mathscr{X}$ under the transformation. The conclusion of Theorem 31 then can be written

$$\int_{\mathscr{R}^k} f(x)\,\mathrm{d}m = |C| \int_{\mathscr{R}^k} f[Ct]\,\mathrm{d}m.$$

In more general cases, particularly when the mapping is not one-to-one, it is essential to bear in mind that the class of functions which can appear as integrands in the integral

$$\int f[\alpha(t)]\,\mathrm{d}\nu$$

is restricted by the nature of $\alpha(t)$. For example, if $\mathcal{T}=\mathcal{X}=\mathcal{R}$ and $\alpha(t)=\sin t$, $f[\alpha(t)]$ is necessarily periodic in $-\infty<t<\infty$. No difficulty arises in any application of the theorem if we start by thinking of $f(x)$ as given over $\mathcal{X}$. Thus, the extension of the last formula to the case in which $x=\alpha(t)$ defines a general differentiable one-to-one mapping of a set T of $\mathcal{R}^k$ onto a set X of $\mathcal{R}^k$ takes the form

$$\int_X f(x)\,dm=\int_T f[\alpha(t)]J(t)\,dm,$$

where $J(t)$ is the Jacobian of $\alpha(t)$. The proof requires more intricate elementary analysis, but no new idea.

2.5 Product measures and multiple integrals

If $\mathcal{X}$ and $\mathcal{Y}$ are two spaces of points x and y, the space of ordered pairs (x, y) is called their **Cartesian product space** and written $\mathcal{X}\times\mathcal{Y}$. The following theorem shows how a **product measure** and integration can be extended to a product space when they have been established in the spaces separately.

Theorem 32. *Suppose that μ, ν are simple measures in rings of (simple) sets I, J in $\mathcal{X}$ and $\mathcal{Y}$. Then the simple sets K in $\mathcal{X}\times\mathcal{Y}$, each consisting of finite union of rectangular sets of type $I\times J$, form a ring on which there is a simple measure m such that $m(K)=\mu(I)\nu(J)$ when $K=I\times J$.*

We define $m(K)$ for a simple set to be the sum of $\mu(I)\nu(J)$ for its component rectangular sets $I\times J$. There is no ambiguity in this as it is easy to verify that the sum is unchanged if the decomposition into rectangular sets is carried out in a different way. It is also easy to see that the simple sets form a ring on which m is finitely additive. All that remains is to show that

$$m(K)=\sum_{v=1}^{\infty} m(K_v)$$

if K and $K_v=I_v\times J_v$ are rectangular sets, $K=\bigcup K_v$ and K_v are disjoint.

Let $\gamma=\gamma(x, y)$, $\gamma_v=\gamma_v(x, y)$ be the characteristic functions of K, K_v, respectively, so that

$$\gamma=\sum_{v=1}^{\infty}\gamma_v.$$

For each v, γ_v is integrable in y for each fixed x and

$$\begin{aligned}\int \gamma_v\,d\nu &= \nu(J_v) && \text{for } x \text{ in } I_v,\\ &=0 && \text{for } x \text{ outside } I_v.\end{aligned}$$

Hence, $\int \gamma_v \, d\nu$ is integrable with respect to μ and

$$\int d\mu \int \gamma_v \, d\nu = \mu(I_v)\nu(J_v) = m(K_v).$$

But it follows from Theorem 19 that $\gamma = \sum \gamma_v$ is also integrable with respect to ν and that

$$\int \gamma \, d\nu = \int \sum \gamma_v \, d\nu = \sum \int \gamma_v \, d\nu.$$

By Theorem 19 again, this is integrable with respect to μ and

$$m(K) = \int d\mu \int \gamma \, d\nu = \int d\mu \sum \int \gamma_v \, d\nu = \sum \int d\mu \int \gamma_v \, d\nu = \sum m(K_v).$$

The function m on simple sets can now be used to establish a measure $m(W)$ over a σ-ring of measurable sets W in $\mathscr{X} \times \mathscr{Y}$ and an associated space of integrable functions. Their integrals over $\mathscr{X} \times \mathscr{Y}$ are generally called **multiple integrals** to distinguish them from the **repeated integrals** already used in the proof of the last theorem. In fact, the two kinds of integrals are closely related as the following theorem shows.

Theorem 33 (Fubini). *If μ and ν are measures in $\mathscr{X}$ and $\mathscr{Y}$ and if $f = f(x, y)$ is measurable with respect to the product measure m of μ and ν, it is measurable with respect to μ and ν for almost all y and x, respectively, and the existence of any one of the integrals*

$$\int |f| \, dm, \qquad \int d\mu \int |f| \, d\nu \qquad \text{or} \qquad \int d\nu \int |f| \, d\mu$$

implies the existence and equality of the integrals

$$\int f \, dm, \qquad \int d\mu \int f \, d\nu, \qquad \int d\nu \int f \, d\mu.$$

We may suppose that $f \geq 0$ and that $f \leq C$ in a simple set K and $f = 0$ in K'. The conclusion in general then follows from Theorem 20 by an argument used in Theorem 30. We can now define simple functions θ_v so that

$$0 \leq \theta_v \leq C, \qquad \theta_v \to f \quad \text{in} \quad K - S, \qquad m(S) = 0.$$

If $\varepsilon > 0$, we can now define rectangular sets in K whose characteristic functions γ_j satisfy

$$\gamma_s \leq \sum \gamma_j, \qquad \sum \int \gamma_j \, dm < \varepsilon.$$

Then

$$\| \| \| \gamma_s \|_\nu \|_\mu \leq \sum \| \| \| \gamma_j \|_\nu \|_\mu = \sum \int \gamma_j \, dm < \varepsilon$$

for every positive ε, and therefore for almost all x, γ_s is null in $\mathscr{Y}$ and $\theta_v \to f$ for almost all y. If we observe that the theorem is obviously true when f is a simple function, the conclusion in general follows from Theorem 18.

2.6 Inequalities and L_p-spaces

The space L_p is defined for $p \geq 1$ as the set of function f for which $|f|^p \in L$, and we write

$$\|f\|_p = \left\{ \int |f|^p \right\}^{1/p}, \qquad \|f\| = \|f\|_1 = \int |f|$$

for any such function, and regard the conditions $f \in L_p$, $|f|^p \in L$ and $\|f\|_p < \infty$ as synonymous. There is no restriction on the space $\mathscr{X}$ on which the integrals are defined, but we indicate it by writing $L_p(\mathscr{X})$ for L_p when this is necessary. We observe also that, since $|f| \leq \max(|f|^p, 1)$, $L \subset L_p$ when $\mathscr{X}$ has finite measure, but that there is no such inclusion if its measure is finite.

If $p > 1$, we define its conjugate p' (and the conjugate space $L_{p'}$) by $p' = p(p-1)^{-1}$, and the limiting case $p' = 1$, $p = \infty$ can be included by defining $\|f\|_\infty$ as the essential upper bound of $|f(x)|$. This means that $\|f\|_\infty$ is the greatest lower bound of numbers α for which $f(x) \leq \alpha$ a.e. It is easy to see, in fact, that $\|f\|_\infty = \lim_{p \to \infty} \|f\|_p$ in a space of finite measure.

The inequalities which follow have wide application in analysis and in application and are indispensible in the study of L_p spaces. The case $p = p' = 2$ is by far the most important and familiar.

Theorem 34 (Hölder's inequality). *If $p \geq 1$, $f \in L_p$, $g \in L_{p'}$, then $fg \in L$ and*

$$\|fg\| \leq \|f\|_p \, \|g\|_{p'}.$$

(The case $p = p' = 2$ is usually called Schwartz's inequality.)

Lemma. *If $1 < p < \infty$ and $\alpha \geq 0$, $\beta \geq 0$, then $\alpha\beta \leq \alpha^p/p + \beta^{p'}/p'$.*

We write $q(t) = 1/p + t^{p'}/p' - t$, so that $q'(t) = t^{p'-1} - 1 < 0$ for $0 \leq t < 1$ and $q'(t) > 0$ for $t > 1$ and $q(t)$ has minimum value 0 when $t = 1$. Then $q(t) \geq 0$ and the conclusion follows on putting

$$t = \beta\alpha^{1-p}.$$

To prove the theorem, we write $a=\|f\|_p$, $b=\|g\|_{p'}$, and take $\alpha=\alpha(x)=|f(x)|\,a^{-1}$, $\beta=\beta(x)=|g(x)|\,b^{-1}$ in the lemma. This gives

$$|f(x)g(x)|\,a^{-1}b^{-1}\leqslant|f(x)|^p\,p^{-1}a^{-p}+|g(x)|^{p'}\,(p')^{-1}b^{-p'}$$

and, on integrating,

$$\|fg\|\,a^{-1}b^{-1}\leqslant p^{-1}+(p')^{-1}=1,$$

as required.

Theorem 35 (Minkowski's inequality). *If* $1\leqslant p\leqslant\infty$ *and* $f_t\in L_p(\mathscr{X})$ *for* $-\infty<t<\infty$ *and G has bounded variation in every finite t-interval, then*

$$\left\|\int f_t\,\mathrm{d}G(t)\right\|_p\leqslant\int\|f_t\|_p\,|\mathrm{d}G(t)|.$$

In particular, $\|f+g\|_p\leqslant\|f\|_p+\|g\|_p$ *for any functions, f, g of* L_p.

We can obviously assume that $G(t)$ increases, that $f_t(x)\geqslant 0$ and that f_t and G are bounded, for the general result then follows from Theorem 20.

For the proof, we write $h(x)=\int f_t(x)\,\mathrm{d}G(t)$,

$$\|h\|_p^p=\int h^p\,\mathrm{d}x=\int h\cdot h^{p-1}\,\mathrm{d}x=\int h^{p-1}\,\mathrm{d}x\int f_t(x)\,\mathrm{d}G(t)$$

$$=\int\left\{\int f_t(x)h^{p-1}\,\mathrm{d}x\right\}\mathrm{d}G(t)\leqslant\int\|f_t\|_p\,\|h\|_p^{p-1}\,\mathrm{d}G(t)$$

by Theorem 34, and the conclusion follows division by $\|h\|_p^{p-1}$.

Minkowski's Theorem in the second and simpler form is an immediate corollary when $G(t)$ is a step function with two steps, and can be interpreted as the 'triangle rule' for the metric $\|f\|_p$. It follows that L_p is a linear metric space in which the 'convergence in L_p' of a sequence f_n to a limit f in L_p means that $\|f_n-f\|_p\to 0$ as $n\to\infty$.

Theorem 36. *If* $p\geqslant 1$, $f_n, f\in L_p$ *and* $f_n\to f$ *in* L_p, *and if* $g\in L_{p'}$, *then*

$$\|f_n\|_p\to\|f\|_p,\qquad \|f_ng-fg\|\to 0,\qquad \int f_ng\to\int fg.$$

For

$$\|f_n\|_p=\|f+f_n-f\|_p\leqslant\|f\|_p+\|f_n-f\|_p,$$

while

$$\|f_n\|_p+\|f-f_n\|_p\geqslant\|f_n+f-f_n\|_p=\|f\|_p$$

and the first part follows. The second follows from Hölder's inequality and Theorem 11, since

$$\left|\int f_n g - \int fg\right| \leq \|f_n g - f_g\| \leq \|f_n - f\|_p \, \|g\|_{p'}.$$

It is important to note that convergence in L_p does not imply convergence a.e. or, indeed, at even a single point.

Another property of L_p which is needed in the sequel is that of **completeness** in the sense of the following theorem.

Theorem 37. *If* $1 \leq p < \infty$ *and* $f_n \in L_p$ *for* $n = 1, 2, \ldots$ *and if* $\|f_n - f_m\|_p \to 0$ *as* $n, m \to \infty$, *then* (i) *there is a function* f *of* L_p *and defined uniquely a.e. such that* $\|f_n - f\|_p \to 0$ *and*

(ii) *there is a subsequence* n_k *for which* $f_{n_k}(x) \to f(x)$ *a.e.*

Choose the sequence n_k so that $\|f_n - f_{n_k}\|_p \leq 2^{-k}$ for $n \geq n_k$. Then

$$\phi_k(x) = \sum_{j=1}^{k-1} |f_{n_{j+1}}(x) - f_{n_j}(x)|$$

increases with k and

$$\|\phi_k\|_p \leq \sum_{j=1}^{k-1} \|f_{n_{j+1}} - f_{n_j}\|_p \leq \sum_{j=1}^{k-1} 2^{-j} < 1.$$

It follows from Theorem 20 that $\phi_k(x)$ tends to a limit $\phi(x)$, that $\phi \in L_p$ and $\phi(x)$ is finite a.e. This means that the series

$$f_{n_1}(x) + \sum_{j=1}^{\infty} [f_{n_{j+1}}(x) - f_{n_j}(x)]$$

converges absolutely and its partial sums $f_{n_k}(x)$ converge a.e. to the values of a function f which is also in L_p since $|f_{n_k}(x)| \leq \phi_k(x) \leq \phi(x)$. It follows then from Theorem 35 that for $n \geq n_k$,

$$\|f_n - f\|_p \leq \|f_n - f_{n_k}\|_p + \|f - f_{n_k}\|_p \leq 2^{-k} + \sum_{j=k}^{\infty} \|f_{n_{j+1}} - f_{n_j}\|_p \leq 2^{-k} + \sum_{j=k}^{\infty} 2^{-j}.$$

Hence $\limsup \|f_n - f\|_p \leq 2^{-k+2}$, and both parts of the conclusion follow on letting $k \to \infty$.

Finally, we show that simple functions are *dense* in L_p.

Theorem 38. *If* $f \in L_p$, $p \geq 1$, *and* $\varepsilon > 0$, *we can define a simple function* θ *so that* $\|f - \theta\|_p < \varepsilon$.

If we define f_n by

$$\begin{aligned} f_n(x) &= f(x) && \text{if} && |f(x)| \leq n, \\ &= n && \text{if} && f(x) \geq n, \\ &= -n && \text{if} && f(x) < -n, \end{aligned}$$

we have $|f_n(x)-f(x)|\leqslant|f(x)|$ and $|f_n(x)-f(x)|\to 0$ a.e., and it follows from Theorem 18 that $\|f_n-f\|_p<\varepsilon/2$ for some value of n. Then if θ is a simple function,

$$\|f-\theta\|_p\leqslant\|f-f_n\|_p+\|\theta-f_n\|_p\leqslant\varepsilon/2+(2n)^{1/p'}\|\theta-f_n\|^{1/p}$$

and the second term can be made less than $\varepsilon/2$ for some θ by the definition of the integral.

3
Integrals of functions of real variables

3.1 Lebesgue and Stieltjes integrals

The general theory developed in Chapter 2 can be applied immediately to the case in which $\mathscr{X}$ is the k-dimensional space $\mathscr{R}^k$ of real vectors $(x_1, x_2, \ldots x_k)$ and $\mu(I)$ is defined in the ring of figures defined in Section 1.4. We are concerned mainly in this chapter with the case $k=1$, $\mathscr{X}=\mathscr{R}$ and the simple sets are then based on intervals $a \leqslant x < b$ and the simple functions are called step functions and take constant values in intervals of this type. We have already seen that a simple measure can be obtained from an increasing point function with values $\mu(x)$, and we assume throughout the chapter that the measure $\mu(X)$ is obtained from such a simple measure by the methods described in Chapter 2. If $\mu(x)$ is not monotonic, but of bounded variation, it still defines a completely additive set function $\mu(X)$ and we can define the integral of a function f with respect to μ as in Section 2.2, so that

$$\int f \, \mathrm{d}\mu = \int f \, \mathrm{d}\mu^+ - \int f \, \mathrm{d}(-\mu^-),$$

where $\mu^+(X)$, $\mu^-(X)$ are the components of $\mu(X)$ corresponding to two monotonic point functions. There is obviously no real restriction in confining the general theory to the monotonic case with $\mu(X) \geqslant 0$, $\mu(x)$ increasing.

The integral $\int f \, \mathrm{d}\mu$ of f with respect to a general increasing μ is called the Lebesgue–Stieltjes integral of f with respect to μ. In the particularly important case $\mu(x)=x$, the integral is called the Lebesgue integral of f (or simply the integral of f) and written $\int f \, \mathrm{d}x$. The integrals can be defined over any set X which is measurable with respect to μ, and when there is any possibility of having to deal with more than one measure, it is convenient to make sure of the necessary measurability conditions by supposing that f is a Borel function and X a Borel set. We show in Theorems 1, 2, and 3 below that there is little loss of generality in this.

We use the familiar notation $\int_a^b f \, \mathrm{d}x$, or sometimes simply $\int_a^b f$, for the integral of f over the interval with end points a and b, and note that it is immaterial whether the end points belong to the interval or not, since the measure of each of them is zero with respect to x. In the Stieltjes case, however, rather more care is needed, since a single point has positive measure with respect to $\mu(X)$ if it is a discontinuity of $\mu(x)$. When this occurs, we distinguish between the four possible cases by writing

$$\int_{a-0}^{b-0}, \int_{a-0}^{b+0}, \int_{a+0}^{b-0}, \int_{a+0}^{b+0}$$

according as the interval is $a \leqslant x < b$, $a \leqslant x \leqslant b$, $a < x < b$, $a < x \leqslant b$, but there is no need for this complication if μ is known to be continuous at a and b, since the four integrals are then all equal.

Since we are interested in this chapter in integrals over the very special space $\mathcal{R}$, it is important to see just how the specific properties of this space are involved in the theory. We have used the topological properties of $\mathcal{R}$ to establish the simple measures $\mu(X)$ over intervals, but when this has been done there is no need to refer to $\mathcal{R}$ throughout the whole of the theory developed in Chapter 2. We return to $\mathcal{R}$ only when we wish to investigate and characterise more closely the classes of measurable sets and integrable functions. Above all, we need to show that the theory applies to the sets and functions which arise naturally in analysis and which are generally classified by their algebraic and topological properties in relation to $\mathcal{R}$. The following theorems show that this can be done very satisfactorily.

Theorem 1. *Open and closed sets are Borel sets and therefore measurable with respect to every μ measure.*

Let X be open and let r be any rational point in X. Let $I(r)$ be the largest interval $r \leqslant x < q$ contained in X. Then if ξ is any point of X, there is an interval $\xi - \delta < x < \xi + \delta$ contained in X, and we can choose a rational r in $\xi - \delta < r < x$ so that $I(r)$ contains x. Since the number of rationals in X is countable, this means that X is the countable union of intervals $I(r)$, and is therefore a Borel set. A closed set, being the complement of an open set, is therefore also a Borel set.

Theorem 2. *A monotonic function f is a Borel function.*

If $f(x)$ increases, the set $\{x : f(x) > c\}$ is an interval of one of the types $a < x < \infty$ or $a \leqslant x < \infty$. The first is open and a Borel set by Theorem 1. The second is the union of the open interval $a < x < \infty$ and the closed set consisting of the single point a.

Theorem 3. *A continuous function f is a Borel function.*

If $X = \{x : f(x) > c\}$ and $\xi \in X$, we can define $\delta > 0$, by the continuity of f at ξ, so that $f(x) > c$ for $\xi - \delta < x < \xi + \delta$ and the whole of this interval belongs to X. This means that X is open for every real c, and the conclusion follows from Theorem 1.

All that need be said now is that the functions of analysis are almost invariably derived from monotonic or continuous functions by algebraic or limit processes which leave intact their Borel properties.

3.2 Some theorems of the integral calculus

Many theorems which are familiar and important in the Riemann Integral Calculus follow immediately for Lebesgue integrals from the results of Chapter 2, and there is no need to state them separately. There are other results, however, which are as familiar as these but which depend essentially on the specific properties of the space $\mathcal{R}$.

This is the appropriate point to clarify the relationship between Riemann and Lebesgue integration. We recall that a bounded function f is R-integrable on $[a, b]$ if it is possible to define step functions ϕ_n, ψ_n so that $\phi_n(x) \leqslant f(x) \leqslant \psi_n(x)$ for $a \leqslant x \leqslant b$ and $\|\psi_n - \phi_n\| \to 0$ as $n \to \infty$; and the Riemann integral defined by $\int f = \lim \int \phi_n$ can then be shown to be unique. As an immediate corollary of Theorem 18 of Chapter 2, we have

Theorem 4. *If f is R-integrable, it is also L-integrable, and the two integrals are equal.*

The condition for R-integrability can now be stated explicitly.

Theorem 5. *A bounded function f is R-integrable in $[a, b]$ if and only if it is continuous a.e. in $[a, b]$.*

We suppose that f is R-integrable and $\varepsilon > 0$, and define a step function ψ so that $f \leqslant \psi$, $\int (\psi - f) \leqslant \varepsilon$. Then if $\eta(\delta, x) = \sup |f(x') - f(x)|$ for $a \leqslant x \leqslant b$, $a \leqslant x' \leqslant b$, $|x' - x| \leqslant \delta$ and if $\psi_\delta(x) = \sup \psi(x')$ for $a \leqslant x \leqslant b$, $a \leqslant x' \leqslant b$, $|x' - x| \leqslant \delta$, we get

$$\eta(\delta, x) \leqslant \psi_\delta(x) - f(x), \qquad \int \eta(\delta, x) \leqslant \int (\psi_\delta - \psi) + \int (\psi - f) \leqslant \int (\psi_\delta - \psi) + \varepsilon.$$

But ψ_δ differs from ψ by less than the upper bound of $|\psi|$ and only in a set of intervals of length δ containing one of the finite set of discontinuities of ψ, and it follows that, as $\delta \to 0$, $\lim \int (\psi_\delta - \psi) = 0$, $\lim \int \eta(\delta, x) \leqslant \varepsilon$. Since ε is as small as we please, and since $\eta(\delta, x)$ decreases with δ, it follows from Theorem 18 of Chapter 2 that $\eta(\delta, x) \to 0$ as $\delta \to 0$ (precisely the condition for the continuity of f at x) holds a.e. Conversely, if f is continuous a.e., then $\int \eta(\delta, x) \to 0$ as $\delta \to 0$ and, for sufficiently small δ, we have $\eta(\delta, x) \leqslant 1/n$ except in a set S of measure $1/n$ or less. If the interval is divided into intervals of length δ, those which contain at least one point of S' have total length at least equal to the measure of S', which is at least $b - a - 1/n$, so that the total length of the remaining intervals is at most $1/n$. The contributions to the difference between upper and lower Riemann sums for the subdivision of $[a, b]$ cannot exceed $(b - a)/n$ and $2 \sup |f|/n$, respectively, and f is therefore R-integrable.

The boundedness of f and of the interval of integration are unavoidable and very serious restrictions on the Riemann theory and are replaced in the Lebesgue theory by the fundamentally lighter condition that f must 'not be too large' in the sense that $\|f\|$ must be finite. The Riemann theory can be extended in this direction only by the introduction of the notion of an 'improper integral'. For example, $\int_a^\infty f(x)\,dx$ can be defined as the limit as $b \to \infty$ of $\int_a^b f(x)\,dx$ when f is R-integrable over every finite interval and the limit exists. Likewise, if f is bounded and integrable in $[a+\delta, b]$ for every $\delta > 0$, but unbounded in $[a, b]$, the integral $\int_a^b f(x)\,dx$ can be defined as the limit as $\delta \to +0$ of $\int_{a+\delta}^b f(x)\,dx$. In fact, of course, the same procedure can be used on L-integrals, but is unnecessary unless $\|f\| = \infty$. It is essential to remember that the general theorems of Lebesgue integration do not hold for improper integrals, whether based on the Riemann or the Lebesgue theory. In these cases, as well as in the Riemann theory generally, limiting processes usually depend on uniformity of convergence and not on the simpler and lighter order of magnitude conditions of Theorems 18, 19, and 20 of Chapter 2. The study of improper integrals is much better regarded as part of the general theory of convergence than of integration in the true sense. In developing the theory for Lebesgue integrals in this section we find that there is little extra trouble in dealing with the general Stieltjes form. The minor difficulties which this generalization causes are due mainly to awkwardness in notation and can be minimised if we remember that μ in the integral $\int f\,d\mu$ represents a measure and not (in this context) a point function. This measure is defined by the function with values $\mu(x-0)$ and is independent of the values of μ at its points of discontinuity. On the other hand, the values of f at its discontinuities may well affect the value of the integral and have to be treated with care. In formulae involving repeated integrals or other complicated expressions, it is often advantageous to write

$$\int d\mu Q \quad \text{instead of} \quad \int Q\,d\mu,$$

as we have already done in Section 2.5. This notation emphasises the fact that $\int$ and μ are the inseparable components of the symbol for the single operation of integration with respect to μ.

Theorem 6 (Integration by parts). *Suppose that F, G are of bounded variation over an interval containing a Borel set X. Then*

$$\int_X F(x-0)\,dG + \int_X G(x+0)\,dF = FG(X),$$

where FG is the set function derived from the point function with values $F(x)G(x)$.

If the values $F(x)$ and $G(x)$ are normalized at common points of *discontinuity so that*

$$2F(x) = F(x+0) + F(x-0), \qquad 2G(x) = G(x+0) + G(x-0)$$

then

$$\int_X F(x)\,\mathrm{d}G + \int_X G(x)\,\mathrm{d}F = FG(X).$$

In particular, if

$$F(x) = \int_a^x f(t)\,\mathrm{d}t, \qquad G(x) = \int_a^x g(t)\,\mathrm{d}t,$$

then

$$\int_a^b F(x)g(x)\,\mathrm{d}x + \int_a^b f(x)G(x)\,\mathrm{d}x = {}_a^b[F(x)G(x)].$$

The three set functions are measures, or sums of at most four measures which are defined uniquely (Theorem 26 of Chapter 2) by their values for intervals, and it is therefore sufficient to prove the theorem in the case when X is the interval I: $a \leqslant x < b$.

The characteristic function of the set $a \leqslant x < b$, $a \leqslant x_1 \leqslant x$ in the product space (x, x_1) is integrable and the two repeated integrals of it with respect to $F(x)$ and $G(x_1)$ are equal, by Fubini's theorem. This gives

$$\int_I [G(x+0) - G(a-0)]\,\mathrm{d}F(x) = \int_I [F(b-0) - F(x_1-0)]\,\mathrm{d}G(x_1)$$

and on replacing x_1 by x, this becomes

$$\begin{aligned}\int_I F(x-0)\,\mathrm{d}G + \int_I G(x+0)\,\mathrm{d}F &= F(b-0)[G(b-0) - G(a-0)] \\ &\quad + G(a-0)[F(b-0) - F(a-0)] \\ &= F(b-0)G(b-0) - F(a-0)G(a-0) \\ &= FG(I).\end{aligned}$$

The second part follows at once if we interchange F and G and add. The last part follows from Theorem 30 of Chapter 2.

Theorem 7 (First mean value theorem). *If f is integrable with respect to μ and $c \leqslant f \leqslant C$ in the interval J: $a \leqslant x \leqslant b$, then*

$$c\mu(J) \leqslant \int_J f\,\mathrm{d}\mu \leqslant C\mu(J),$$

and if $f(x)$ is continuous in $a \leqslant x \leqslant b$, then $\int_J f\,\mathrm{d}\mu = f(\xi)\mu(J)$ for some value of ξ in $a \leqslant \xi \leqslant b$.

The first part is a special case of Theorem 14 of Chapter 2 when J is replaced by a simple interval $a \leqslant x < b+1/n$, and the conclusion follows from the complete additivity of $\mu(X)$ and $\int_X f\,d\mu$ as $n \to \infty$. The second part follows from the fact that a continuous function in a closed interval attains every value between its upper and lower bounds.

Theorem 8 (Second mean-value theorem). (i) *If f is integrable and ϕ monotonic in $a \leqslant x \leqslant b$, then*

$$\int_a^b f\phi\,dx = \phi(a)\int_a^\xi f\,dx + \phi(b)\int_\xi^b f\,dx$$

for some ξ in $a \leqslant \xi \leqslant b$.

(ii) *If ϕ decreases and $\phi \geqslant 0$, then*

$$\int_a^b f\phi\,dx = \phi(a)\int_a^\xi f\,dx$$

for some ξ in $a \leqslant \xi \leqslant b$.

We suppose that ϕ decreases and define $f(x)=0$ outside $a \leqslant x \leqslant b$ and $\phi(a-0)=\phi(a)$, $\phi(b+0)=\phi(b)$ and note that a similar proof applies if ϕ increases. Then if

$$F(x) = \int_a^x f(t)\,dt,$$

it follows from Theorem 6 applied to the interval $a \leqslant x \leqslant b$ and Theorem 30 of Chapter 2 that

$$\begin{aligned}\int_a^b f\phi\,dx &= {}_{a-0}^{b+0}[F\phi] - \int_{a-0}^{b+0} F\,d\phi\\ &= \phi(b)\int_a^b f\,dt - [\phi(a)-\phi(b)]F(\xi)\end{aligned}$$

for some ξ in $a \leqslant \xi \leqslant b$ by Theorem 6. This gives (i), and we deduce (ii) immediately by redefining $\phi(b)=0$, since this change at one point does not change the first two terms in (i).

Theorem 9 (Differentiation under integral sign). *Suppose that $f(x, t)$ is integrable with respect to x in a neighbourhood of t, that*

$$|f(x, t+h) - f(x, t)| \leqslant |h|\,\lambda(x),\ \lambda \in L,$$

and that $\partial f/\partial t$ exists a.e. Then $\partial f/\partial t \in L$ and

$$\int \frac{\partial f}{\partial t}\,dx = \frac{d}{dt}\int f(x, t)\,dt.$$

This follows from the continuous parameter case of Theorem 18 of Chapter 2.

Theorem 10. (i) *If* $f \in L_p(\mathscr{R})$, $p \geq 1$, $\varepsilon > 0$, *we can define a step function* θ *and a continuous function* ϕ *so that* $\|f-\theta\|_p \leq \varepsilon$, $\|f-\phi\|_p \leq \varepsilon$. *If* $f \in L_p(a, b)$, $-\infty < a < p < \infty$, *we can define a polynomial* ψ *so that* $\|f-\psi\|_p < \varepsilon$.

(ii) $\|f(x+\delta)-f(x)\|_p \to 0$ as $\delta \to 0$.

The first part follows immediately from Theorem 38 of Chapter 2, since a simple function in $\mathscr{R}$ is a step function. For the second part, the required continuous function ϕ can be obtained from θ by changing its value in arbitrarily small intervals near its discontinuities. The next part follows from the theorem that a continuous function in a finite interval can be approximated uniformly and so, *a fortiori*, in L_p, by a polynomial. The last part follows from the facts that $\|f(x+\delta)-f(x)\|_p \leq 2\varepsilon + \|\theta(x+\delta)-\theta(x)\|_p$ by Theorem 35 of Chapter 2, and that $\|\theta(x+\delta)-\theta(x)\|_p \to 0$ for a step function θ.

The theory of integration described in Chapter 2 does not depend in any way on the notion of differentiation. In fact, there may not be any natural and simple way of defining the derivative of a set function $F(X)$ unless more is known about the topological properties of the space $\mathscr{X}$. The space $\mathscr{R}$, however, has a very rich and familiar topology and a fully developed theory of differentiation, and the close relationship between integration and differentiation is brought out by the fundamental theorem of calculus, that

$$F(x) = \int_a^x f(t)\, dt$$

is differentiable at every continuity point of the Riemann integrable function f. The following theorem is equally fundamental in the Lebesgue theory and, together with the Radon–Nikodym theorem, shows how integration and differentiation can be treated as inverse operations. It shows in fact, that the Radon–Nikodym derivative as defined in Theorem 28 of Chapter 2 is the familiar differential coefficient in the space $\mathscr{R}$.

Theorem 11 (Fundamental theorem of calculus). (i) *If* $f \in L(a, b)$, *then*

$$\lim_{h\to 0} \frac{1}{h} \int_0^h |f(x+t)-f(x)|\, dt = 0 \text{ a.e. in } (a, b)$$

and $F(x) = \int_a^x f\, dt$ *is differentiable and* $F'(x) = f(x)$ *a.e. in* (a, b)

(ii) *If* $F(x)$ *is absolutely continuous, it is differentiable a.e.,* $F'(x)$ *is integrable and*

$$F(x) = \int_a^x F'(t)\, dt.$$

We make the trivial simplification that $a=0$, $b=1$ and suppose that $g \geqslant 0$, $g \in L(0,1)$ $G(x)=\int_a^x g\, dt$. We then define the value of the step function $g_n(x)$ in each binary interval $v2^{-n} \leqslant x < (v+1)2^{-n}$ $(n \geqslant 1,\ v \geqslant 0)$ to be

$$2^n \int_{v2^{-n}}^{(v+1)2^{-n}} g\, dt.$$

The upper derivative D^+G of G is defined in the usual way by

$$D^+G(x) = \limsup_{h \to 0}[G(x+h)-G(x)]/h = \limsup_{h \to 0} h^{-1} \int_0^h g(x+t)\, dt.$$

If $c>0$, the set E_n in which $g_n(x) \geqslant c$ is the union of binary intervals of length 2^{-n}; and if $2^{-n-1} < |h| \leqslant 2^{-n}$,

$$h^{-1} \int_0^h g(x+t)\, dt \leqslant h^{-1}2^{-n}[g_n(x)+g_n(x-2^{-n})+g_n(x+2^{-n})] \leqslant 6c \quad (1)$$

unless x or $x \pm 2^{-n}$ belongs to E_n, when x must belong to the set E_n^* consisting of the intervals of E_n together with the adjoining intervals of length 2^{-n}. It follows that the set in which $h^{-1}\int_0^h g(x+t)\, dt > 6c$ for some value of h in $2^{-N-1} < h \leqslant 2^{-1}$ is contained in $\bigcup_{n=1}^{N} E_n^*$. This set can be obtained from $\bigcup_{n=1}^{N} E_n$ by replacing each disjoint interval of the latter by an interval three times as long (or less when there is overlapping), and therefore

$$\mu\left(\bigcup_{n=1}^{N} E_n^*\right) \leqslant 3\mu\left(\bigcup_{n=1}^{N} E_n\right), \quad (2)$$

where μ denotes Lebesgue measure.

Since $\bigcup_{n=1}^{N} E_n$ can be expressed as the union of disjoint intervals in each of which the mean value of $g(t)$ is at least c, it follows that

$$c\mu\left(\bigcup_{n=1}^{N} E_n\right) \leqslant \int_a^b g\, dt$$

and since this holds for all N, it follows that

$$c\mu\{x : D^+G > 6c\} \leqslant 3\int_a^b g\, dt. \quad (3)$$

Now let $\varepsilon > 0$ and define a step function $\theta(t)$ so that

$$\int_a^b g\, dt < \varepsilon, \qquad g = |f-\theta|. \quad (4)$$

Then

$$h^{-1}\int_0^h |f(x+t)-f(x)|\,dt$$
$$\leqslant h^{-1}\int_0^h g(x+t)\,dt+g(x)+h^{-1}\int_0^h |\theta(x+t)-\theta(x)|\,dt$$

and, since the last term tends to 0 with h unless x is one of the finite number of discontinuities of $\theta(t)$, we get

$$\limsup_{h\to 0} h^{-1}\int_0^h |f(x+t)-f(x)|\,dt\leqslant D^+G+g\leqslant 7c, \tag{5}$$

by (3) and (4), except in a set of measure at most $4\varepsilon/c$, since, $D^+G\leqslant 6c$ except in a set of measure at most $3\varepsilon/c$, and $g(x)\leqslant c$ except in a set of measure at most ε/c. Since ε may be arbitrarily small, it follows that (5) holds a.e., and, since this is true for every positive c, we have the first part of (i). The second part then follows from the inequality

$$\left|\int_0^h f(x+t)\,dt-hf(x)\right|\leqslant\int_0^h |f(x+h)-f(x)|\,dt,$$

and (ii) is a corollary of this and Theorem 28 of Chapter 2.

The last theorem can be extended to include functions which are not necessarily absolutely continuous, but only of bounded variation.

Theorem 12. *If Q is a singular function of bounded variation, $Q'(x)=0$ a.e. If H is a function of bounded variation with decomposition $H=F+Q$, with absolutely continuous F and singular Q, then $H'(x)=F'(x)$ a.e. If H increases, then*

$$H(I)\geqslant\int_I H'=\int_I F',$$

with equality only if $Q(I)=0$.

After Theorem 12, it is sufficient to prove that $Q'(x)=0$ a.e. if Q is singular and increasing, and this needs only a slight modification in the proof. If we define X to be the set in which

$$D^+Q>6c,$$

the argument leading up to the inequality (3) in the proof of Theorem 12, but with Q replacing G, gives $\mu(XI)\leqslant 3c^{-1}Q(I)$ for every interval I, and it follows from this that $\mu(X\cap Y)\leqslant 3c^{-1}Q(Y)$ for any Borel set Y. But Q is singular, and there is a null Borel set S such that $Q(S')=0$, and on taking $Y=S'$ we get

$$\mu(X)=\mu(X\cap S)+\mu(X\cap S')=\mu(X\cap S')\leqslant 3c^{-1}Q(S')=0,$$

and so $0 \leqslant D^{+}G \leqslant 6c$ a.e. Since this holds for all $c>0$, we have $Q'=0$ a.e.

These results can be used to establish the general forms of the theorems for change of variable.

Theorem 13 (Change of variable). *If $\alpha(t)$ is strictly increasing and continuous for $A \leqslant t \leqslant B$ and $\alpha(A)=a$, $\alpha(B)=b$, and if $G(x)$ is of bounded variation in $a \leqslant x \leqslant b$, then*

$$\int_{a-0}^{b+0} f(x)\,\mathrm{d}G(x) = \int_{A-0}^{B+0} f[\alpha(t)]\,\mathrm{d}G[\alpha(t)]$$

In particular,

$$\int_{a}^{b} f(x)\,\mathrm{d}x = \int_{A}^{B} f[\alpha(t)]\,\mathrm{d}\alpha(t)$$

These conclusions follow from Theorems 30 and 31 of Chapter 2 (without using Theorem 11).

Theorem 14. *If $\alpha(t)$ is absolutely continuous but not necessarily monotonic for $A \leqslant t \leqslant B$ and $\alpha(A)=a$, $\alpha(B)=b$, then*

$$\int_{a}^{b} f(x)\,\mathrm{d}x = \int_{A}^{B} f[\alpha(t)]\alpha'(t)\,\mathrm{d}t.$$

This follows at once from Theorem 11 and the chain rule for differentiation if we observe that the two sides have the same derivative with respect to B.

3.3 Integrals in $\mathscr{R}^k$

Most of the integral calculus developed in the light of the specific properties of R extends readily to $\mathscr{R}^k$ when we make use of the fact that Lebesgue measure m in $\mathscr{R}^k$ is the product measure, in the sense of Section 2.5, of k Lebesque measures over $\mathscr{R}$. The basic tool is Fubini's theorem (Theorem 33 of Chapter 2) which expresses integrals in $\mathscr{R}^k$ as repeated integrals with respect to a single real variable. This also provides a completely satisfactory treatment of the geometrical concepts of area and volume.

Thus a set of points in the plane is said to have an area if and only if it is measurable, and its area is defined simply as its measure. Similarly, a measurable set in $\mathscr{R}^3$ has volume equal to its measure. The following theorems are immediate corollaries of Fubini's theorem.

Theorem 15. *The set of points $\{x, y : 0 \leqslant y \leqslant f(x)\}$ in $\mathscr{R}^2$ has an area, and this area is $\int f\,\mathrm{d}x$, if and only if $f \in L(\mathscr{R})$.*

Theorem 16. *The set of points* $\{x, y, z : 0 \leqslant z \leqslant f(x, y)\}$ *in* $\mathscr{R}^3$ *has a volume, and this volume is* $\iint f(x, y)\,dx\,dy$, *if and only if* $f \in L(\mathscr{R}^2)$.

Both theorems express the fact that an integral over a space $\mathscr{R}^k$ can be used to specify a measure in $\mathscr{R}^{k+1}$. They express areas (and volumes) in terms of boundary curves (and surfaces) of regions under consideration and, as we see later, this is a characteristic which they share with the deeper theorems of Green and Gauss which can be thought of as generalizations of them. The sets in both theorems are restricted by being bounded in part by lines or planes, but this can be dealt with in practice by dividing more complex regions into subregions of this kind. Alternatively, it may be more convenient to apply Fubini's theorem directly and obtain the area of a plane set X, for example, as $\int \mu(x)\,dx$, where $\mu(x)$ is the measure in $\mathscr{R}$ of the intersection of X with the x-ordinate; and the value of a region Z in $\mathscr{R}^3$ as $\int A(z)\,dz$, where $A(z)$ is the area of the intersection of Z with the horizontal plane through $(0, 0, z)$. There is no need for regions and the function defining them to be bounded. The regions

$$\{x, y : x \geqslant 1, 0 \leqslant y \leqslant x^{-2}\} \qquad \text{and} \qquad \{x, y : 0 \leqslant x \leqslant 1, 1 \leqslant y \leqslant x^{-1/2}\}$$

both have area 1, while the area of $\{x, y : x \geqslant 1, 0 \leqslant y \leqslant x^{-1}\}$ is not defined (except in some circumstances, and with due caution, as $+\infty$).

The fundamental theorem of the calculus (Theorem 11) establishes the connection between integration and measure theory and the concept of differentiation defined in relation to the topological properties of $\mathscr{R}$. The same kind of relationship holds in $\mathscr{R}^k$ and is established in a similar way. The starting point is the definition of the derivative of a completely additive set function F with values $F(X)$ over measurable sets X of $\mathscr{R}^k$. We say that F is **differentiable,** with **derivative** $F'(x) = f(x)$, at x if

$$\lim_{r \to 0} F[X(r)]/\mu[X(r)] = f(x) \tag{1}$$

provided only that $X(r)$ is measurable for sufficiently small r and lies in a sphere of radius r with centre x and

$$\liminf r^{-2}\mu[X(r)] > 0.$$

This definition of differentiability is more stringent in the case $k = 1$ than the usual one adopted in Theorem 11, and the following theorem appears to be stronger than the direct analogue of Theorem 11 in which the sets $X(r)$ would be rectangles containing x. The difference is immaterial however, for the slightly stronger version for $k = 1$ would follow equally well from part (i) of Theorem 11.

Theorem 17 (Fundamental theorem of calculus in $\mathscr{R}^k$). (i) *if* $f \in L(\mathscr{R}^k)$, *then*

$$\lim_{r\to 0} r^{-k}\int_{|t|\leq r} |f(x+t)-f(x)|\,\mathrm{d}t = 0 \quad a.e.,$$

and

$$F(x)=\int_X f(x)\,\mathrm{d}x \text{ is differentiable and } F'(x)=f(x) \quad a.e.$$

(ii) *If F is absolutely continuous over the measurable sets X of $\mathscr{R}^k$, then it is differentiable a.e. and*

$$F(X)=\int_X f(x)\,\mathrm{d}x,\ f(x)=F'(x) \quad a.e.$$

The proof of Theorem 11 applies with straightforward modifications, and the same is true for the following extension of Theorem 12

Theorem 18. *If $Q(X)$ is a bounded singular set function in $\mathscr{R}^k$, then $Q'(x)=0$ a.e.*

These last theorems enable us to deal with a problem of great practical importance—the extension to $\mathscr{R}^k$ of the formulae for change of variable stated for $\mathscr{R}$ in Theorems 13 and 14. The extension needs some care since a transformation α on $\mathscr{R}^k$ when $k>1$ does not take a rectangle into a rectangle as is the case when $k=1$ even when α is not strictly one-to-one, and the proof of Theorem 14 has no analogue in $\mathscr{R}^k$. Moreover, although an extension of Theorem 13 to the case $k>1$ can be formulated, it does not add anything to what can be obtained from Theorem 31 of Chapter 2. We therefore confine ourselves to the most important case, which is the direct extension of Theorem 14 under the stronger condition that α is one-to-one. For this we need some results on the differentiation of point and set functions to give an analogue in $\mathscr{R}^k$ of the chain rule for differentiation in $\mathscr{R}$.

A function α on $\mathscr{R}^k$ to $\mathscr{R}^j$ is said to be **differentiable** at t if

$$\alpha(t+h)-\alpha(t)=\alpha'(t)h+o(|h|) \qquad \text{as} \qquad h\to 0$$

and $\alpha'(t)$ is the matrix of values at t of the partial derivatives of α. The chain rule for such transformation is that the composite function γ defined for two differentiable functions α and β by $\gamma(t)=\beta[\alpha(t)]$ is also differentiable and $\gamma'(t)=\beta'[\alpha(t)]\alpha'(t)$. The matrices α, β must obviously be comformable but need not necessarily be square. However, if α is square, the determinant J (or J_α) of α' is called the **Jacobian** of α and clearly $J_{\beta\alpha}=J_\beta J_\alpha$ when α,β are both square. The validity of this chain rule is clear from the elementary theory of partial differentiation.

When $k=j$, the function α also defines a transformation of sets in which any set X in $\mathscr{R}^k$ is taken into its image $\alpha(X)$ in $\mathscr{R}^k$, and if α is also one-to-one and F is a completely additive set function, then F_α defined by $F_\alpha(X)=F[\alpha(X)]$ is also completely additive. The most important case is that in which $F(X)$ is the ordinary Lebesgue measure $\mu(x)$ of X and the absolute continuity of μ_α is established, after Theorem 27 of Chapter 2, if $\mu_\alpha(X)=0$ for every null set X.

The following theorem gives another extension of the same familiar chain rule, but with the novel feature that one function is a point function while the other is a set function.

Theorem 19. *Suppose that α is a one-to-one differentiable function on an open region of $\mathscr{R}^k$ in which μ_α is absolutely continuous and that the set function F is absolutely continuous in the image region. Then F_α is also absolutely continuous and has derivative $f[\alpha(t)]\ J(t)$ a.e.*

Corollary. *For any such function α, $J(t)\geqslant 0$ a.e.*

If $\varepsilon>0$, we can choose δ_1 so that $|F(X)|\leqslant\varepsilon$ when $\mu(X)\leqslant\delta_1$, and then choose δ_2 so that $\mu[\alpha(X)]\leqslant\delta_1$ when $\mu(X)\leqslant\delta_2$, and this is enough to show that F_α is absolutely continuous. It is therefore differentiable a.e. by Theorem 17 and its derivative is given by any appropriate choice of sets $X(r)$, $r>0$, in the formula (1) above.

We take $X(r)=I(r)$, the cube with centre t and side $2r$ and suppose that t is a point at which α is differentiable. Then $\alpha(I)$ lies within a sphere of radius Br for some constant B and sufficiently small r since $|h|^{-1}\,|\alpha(t+h)-\alpha(t)|$ is bounded by a number depending on the (finite) values of the terms of J. The differentiability of α ensures that $\alpha(I)$ both contains and is contained in parallelepipeds $\alpha_1(I)$ and $\alpha_2(I)$ which satisfy the two conditions that they lie within a sphere of radius $2Br$ for sufficiently small r and also that

$$\mu[\alpha,(I)]=J(t)[\mu(I)+o(1)],\ \mu[\alpha_2(I)]=J(t)[\mu(I)+o(1)] \qquad \text{as} \qquad r\to 0$$

(We omit some straightforward but tedious details of the coordinate geometry of the parallelepiped in $\mathscr{R}^k$). It follows that

$$\mu[\alpha(I)]=J(t)[\mu(I)+o(1)]$$

and hence

$$\begin{aligned} F_\alpha(I)=F[\alpha(I)]&=\mu[\alpha(I)]\{f[\alpha(t)]+o(1)\} \\ &=\mu(I)\{J(t)+o(1)\}\{f[\alpha(t)]+o(1)\} \end{aligned}$$

and this is enough to show that $F'_\alpha(t)=J(t)\ f[\alpha(t)]$ a.e.

The corollary follows at once if we put $F(X)=\mu(X)$.

Theorem 20 (Change of variable). *Suppose that α is a one-to-one differentiable function on an open region of $\mathcal{R}^k$ in which μ_α is absolutely continuous and that f is measurable. Then*

$$\int_{\alpha(T)} f(x)\,\mathrm{d}x = \int_T f[\alpha(t)]J(t)\,\mathrm{d}t$$

in the sense that if either integral exists, so does the other, and the two are equal.

We note from the last theorem that the set function F_α defined by

$$F_\alpha(T) = \int_{\alpha(T)} f(x)\,\mathrm{d}x$$

is absolutely continuous and

$$F_\alpha = f[\alpha(t)]J(t),$$

and the conclusion then follows from the fundamental theorem (Theorem 17).

PART II

4
Measure and integration in geometry

4.1 Areas, volumes, and mass

We have seen in Chapter 3 that the familar notions of area in $\mathscr{R}^2$ and volume in $\mathscr{R}^3$ can be taken as synonymous with Lebesgue measure in these spaces. An immediate generalization arises if we replace these by more general measures, for we can interpret any such measure as a distribution of mass. If the measure of a set X in a mass distribution is denoted by $M(X)$, the Radon–Nikodym theorem shows that $M(X)=D(X)+Q(X)$, where D is absolutely continuous with respect to Lebesgue measure μ (volume in $\mathscr{R}^3$, area in $\mathscr{R}^2$) and $Q(X)=Q(X\cap S)$, where $\mu(S)=0$. Then D has derivative $d(x)$ at almost all points x, and

$$D(X)=\int_X d(x)\,\mathrm{d}\mu.$$

If $Q=0$, we say that M defines a **density distribution** with **density** $d(x)$. On the other hand, if $D=0$, the whole mass is concentrated in a set of S of measure zero and, if S consists of a countable set of points x_i, we have a point mass distribution consisting of point masses m_i at x_i. Strictly speaking, of course, the concept of density is inconsistent with the atomic theory of matter, but the errors introduced by using mathematical models based on density are rarely significant.

Some very important and familiar physical quantities can be expressed very naturally as Stieltjes integrals of functions on $\mathscr{R}^k$ with respect to a mass distribution. First, the mean centre, or centre of mass $\bar{r}$ of a distribution M is given in vector form by

$$\bar{r}=M^{-1}\int r\,\mathrm{d}M,$$

where $M=\int \mathrm{d}M$.

A related idea in $\mathscr{R}^3$ is the **moment of inertia** I of a mass distribution about a line. It is defined by

$$I=\int p^2\,\mathrm{d}M,$$

where p is the distance of a point r in $\mathscr{R}^3$ from the line. It is easy to derive the important fact that

$$I=I_0+Ma^2,$$

where a is the distance of the line from the centre of mass and I_0 is the moment of inertia of the distribution about the parallel line through this centre. It follows, of course, that the moment of inertia about parallel lines is least for the line through the mass centre.

The idea of a distribution defined by a measure over $\mathscr{R}^2$ or $\mathscr{R}^3$ obviously extends to physical quantities other than mass (electric charge, for example) and, as we see in more detail in Chapter 6, to the basic concepts of probability.

4.2 Curves

A (parametrized) **curve** in $\mathscr{R}^k$ is the image in $\mathscr{R}^k$ of a closed interval $[a, b]$ under a continuous function on $[a, b]$ into $\mathscr{R}^k$. The value of k is immaterial in most of this section, but visualization is easiest and most useful in the case $\mathscr{R}^3$, and we then use the vector notation

$$r(t)=[x(t), y(t), z(t)], \qquad a \leqslant t \leqslant b,$$

to indicate 'the point t' on the curve, t being called the **parameter.** There is no loss of generality in assuming, as we always do, that $r(t)$ does not remain constant in any interval of values of t.

A curve between two points is sometimes called an *arc* simply to distinguish it from its *chord,* which is the straight line joining its end points. A curve is called **simple** if

$$r(t) \neq r(t') \quad \text{for} \quad a \leqslant t < t' \leqslant b$$

and closed if $r(a)=r(b)$. It is called continuous if r is continuous and differentiable if r is differentiable at a single point or in the interval, and other properties of a curve can be expressed in the same way. A common condition is that a curve is piecewise continuously differentiable, which means that $[a, b]$ can be divided into a finite number of segments inside each of which r is differentiable with continuous derivative, including the appropriate left or right derivatives at the common ends of segments. Such curves include polygonal curves and other curves with well defined corners and, when the curve is closed, it is assumed that the behaviour of the curve at and near the point

$$r(a)=r(b)$$

is appropriate. In fact, it is often useful to think of a closed curve as the image of the unit circle under a continuous mapping into $\mathscr{R}^k$. The condition of piecewise continuous differentiability is a reasonable one in many applications, but it is not the most general or the most natural in this particular context, and we find that absolute continuity is more appropriate to the spirit of the last chapter.

If a curve is regarded primarily as a set of points in $\mathscr{R}^k$, the parametrization of it is largely arbitrary. For if the function τ defined by $t \to \tau(t)$ on the interval $a \leq t \leq b$ is strictly increasing and continuous, then

$$r^*(t) = r[\tau(t)], \qquad a \leq t \leq b,$$

defines an alternative parametrization on

$$\tau(a) \leq \tau \leq \tau(b)$$

of the curve which is geometrically identical with that described by r. The same is true, of course, for a simple closed curve parametrized as an image of the unit circle. Two special cases are important. The first is that in which the parameter t can be interpreted as a measure of time, for we then have a very simple and powerful way of describing the motion of a particle along a curve, with the immediate kinematic ideas of **velocity** $r'(t)$, **acceleration** $r''(t)$ and **speed** $|r'(t)|$. The second is the case in which the parameter is the **arc length** $s(t)$ of the curve from a to t, but we have first to define this. The most satisfying approach is that derived from the intuitive geometrical idea of the length of a curve as the limit, in some sense, of the lengths of inscribed polygonal curves as they approach it. Such an inscribed polygonal curve P is defined by taking a finite number of points

$$r_j = r(t_j), \qquad a = t_0 < t_j < t_{j+1} < t_J = b$$

on the curve C and joining them consecutively by linear segments. The length $s(P)$ of the polygonal curve is given by elementary geometry as

$$s(P) = \sum_{j=1}^{J} |r_j - r_{j-1}|$$

and we say that C is **rectifiable** if $s(P)$ is bounded for all P, and define its length $s(a, b)$ [or $s(C; a, b)$] as the least upper bound of $s(P)$. It is easy to check that s is finitely additive in the sense that $s(a, c) = s(a, b) + s(b, c)$ when $a < b < c$, for it follows from the fact that the sum of lengths of two sides of a triangle is greater than the third that the length of a polygonal curve is not decreased by the insertion of an extra vertex. Thus, given $\varepsilon > 0$, we can define P so that

$$s(P; a, b) \geq s(a, b) - \varepsilon, \qquad s(P; b, c) \geq s(b, c) - \varepsilon, \qquad s(P; a, c) \geq s(a, c) - \varepsilon,$$

and it follows that

$$\begin{aligned} s(a, c) \geq s(P; a, c) &= s(P; a, b) + s(P, b, c) \geq s(a, b) + s(b, c) - 2\varepsilon \\ &\geq s(P; a, b) + s(P; b, c) - 2\varepsilon = s(P; a, c) - 2\varepsilon \\ &\geq s(a, c) - 3\varepsilon, \end{aligned}$$

and this is sufficient to establish additivity and the fact that $s(t)=s(a,t)$ is strictly increasing in $[a,b]$.

The basic result in this section is

Theorem 1. *If C is an absolutely continuous curve defined by the mapping $t \to r(t)$ for $a \leqslant t \leqslant b$ (with r absolutely continuous), then s is absolutely continuous, $s'(t)=|r'(t)|$ a.e. and*

$$s(t)=s(a,t)=\int_a^t |r'| \qquad \text{for} \qquad a \leqslant t \leqslant b.$$

If the vertices of a polygonal curve joining $r(a)$ and $r(t)$ are $r_j = r(t_j)$, we have

$$r_j - r_{j-1} = \int_{t_{j-1}}^{t_j} r'$$

by Theorem 11 of Chapter 3, since r is absolutely continuous and so, using Theorem 11 of Chapter 2,

$$\sum |r_j - r_{j-1}| = \sum \left| \int_{t_{j-1}}^{t_j} r' \right| \leqslant \sum \int_{t_{j-1}}^{t_j} |r'| = \int_a^t |r'|,$$

and since this holds for all polygonal curves inscribed in C, it follows that

$$s(t) \leqslant \int_a^t |r'|.$$

The same argument applied to all sub-intervals of $[a,b]$ shows that s is also absolutely continuous.

On the other hand, it follows from the definition of $s(t)$ and the fact that the chord joining two points of C is itself a polygonal curve, that

$$h^{-1}[s(u+h)-s(u)] \geqslant h^{-1}|r(u+h)-r(u)| \quad \text{if} \quad h>0$$

and that $s'(u) \geqslant |r'(u)|$ at points u where both $s'(u)$, $r'(u)$ exist, and this is almost everywhere. It follows now from Theorem 11 of Chapter 3 that

$$s(t)=\int_a^t s' \geqslant \int_a^t |r'|$$

and, with the opposite inequality already proved, that

$$s(t)=\int_a^t |r'| \quad \text{and} \quad s'(t)=|r'| \quad \text{a.e.}$$

In the more general case of a curve which is rectifiable, but not necessarily absolutely continuous, the arc length $s(t)$ is still strictly increasing and differentiable a.e., but instead of the full conclusion of

Theorem 1, we can only say that

$$s(t) \geq \int_a^t |r'|,$$

with equality only when r is absolutely continuous. An example in $\mathscr{R}^2$ is a curve $r = (x, y)$ with $x(t) = t$ and y continuous and singular and increasing strictly from 0 to 1 in $0 \leq t \leq 1$. This must have length s at least $2^{1/2}$, while $x' = 1$, $y' = 0$ and $\int |r'| = 1 < s$. Such examples are rarely of interest in application and little is lost by confining the theory to absolutely continuous rectifiable curves.

If f is a scalar function defined on a region of $\mathscr{R}^3$ which contains an absolutely continuous curve C and is such that the function with values $f[r(t)]$ is measurable in $a \leq t \leq b$, we can define **integrals of** f **along** C by

$$\int_C f \, dx = \int_a^b f[r(t)]x'(t) \, dt$$

and $\int_C f \, dy$, $\int_C f \, dt$ similarly, and

$$\int_C f \, ds = \int_a^b f[r(t)]s'(t) \, dt.$$

If f is a vector function defined in the same way on C, we define

$$\int_C f \cdot ds = \int_a^b f[r(t)] \cdot r'(t) \, dt.$$

This last integral is familiar in mechanics as the work done by a force f in moving a particle from $r(a)$ to $r(b)$ along the curve. A vector valued function f defined in an open region is often called a vector field and is called **conservative** if $\int f \cdot ds$ is the same when taken along all rectifiable curves joining two points in the region. If f is the gradient (grad ϕ) of some potential function defined in the region, then

$$\int_C f \cdot ds = \int_a^b \left(\frac{\partial \phi}{\partial x} \frac{dx}{dt} + \frac{\partial \phi}{\partial y} \frac{dy}{dt} + \frac{\partial \phi}{\partial z} \frac{dz}{dt} \right) dt = \int_a^b \frac{\partial \phi}{\partial t} \, dt = \phi[r(b)] - \phi[r(a)].$$

Conversely, if $f = (p, q, w)$ is integrable along every absolutely continuous curve in the region and if

$$\int_C f \cdot ds$$

is the same for all such curves from a fixed point to a variable point r, we define $\phi(r)$ to be this common value. Then if i is the unit vector $(1, 0, 0)$, we have

$$\phi(r+hi)-\phi(r)=\int_0^h p(r+ti)\,dt$$

$$\frac{\partial\phi}{\partial x}=\lim_{h\to 0}\frac{\phi(r+hi)-\phi(r)}{h}=\lim_{h\to 0}\frac{1}{h}\int_0^h p(r+ti)\,dt=p$$

for almost all r and, in particular, at any point of continuity of p. Similarly,

$$\frac{\partial\phi}{\partial y}=q, \qquad \frac{\partial\phi}{\partial z}=r \quad \text{and} \quad f=\operatorname{grad}\phi \quad \text{a.e.}$$

4.3 Surfaces

A parametrized **surface** D in $\mathscr{R}^3$ with boundary curve C is defined, by analogy with the previous definition of a curve, as the image in $\mathscr{R}^3$ of a closed domain Δ and its boundary Γ in $\mathscr{R}^2$ under a function which is continuous on $\Delta\cup\Gamma$. The surface is called **simple** if the mapping is one-to-one between $\Delta\cup\Gamma$ and $\mathrm{D}\cup\mathrm{C}$. Except in cases when we have to consider more than one domain Δ, there is generally no loss of generality in supposing that Δ is the unit disk and Γ its boundary circle. This definition excludes surfaces which are 'closed' in the geometrical sense in which the surface of an ellipsoid, for example, is closed, but we can include them by defining a simple closed surface D and the region L of $\mathscr{R}^3$ contained in and on it as the image in $\mathscr{R}^3$, under a one-to-one continuous function, of the surface Δ of the unit sphere and the union Λ of Δ and its interior. A closed surface has no boundary, of course, but we can still consider the boundaries of suitably defined parts of it.

To complete the analogy with curves, it would be necessary to relate these definitions to the concept of a surface as the limit of inscribed polyhedral surfaces whose geometrical properties are familiar. This can be done and there is an extensive literature on it, but there are real topological difficulties in any treatment which is even minimally satisfactory. There is no space for such a theory in a book of this size and it would in any case take us too far from its central themes. We therefore omit it entirely except to note that the following analysis is valid for polyhedral surfaces themselves and leads to familiar and expected conclusions when applied to them.

The parametric equations for a surface can be given in vector form

$$r(u,v)=[x(u,v),\,y(u,v),\,z(u,v)]$$

where the functions x, y, z are continuous and defined for values of the parameters (u, v) on the unit disk or sphere according as we are dealing with open or closed surfaces. The surface is called differentiable if r is differentiable in the sense that partial derivatives r_u, r_v of r with respect to u, v, respectively, exist and

$$\delta r = r_u \delta u + r_v \delta v + o(|\delta u| + |\delta v|) \quad \text{as} \quad \delta u, \delta v \to 0$$

at a point, at all points or at almost all points as is appropriate. Differentiability at a point is sufficient to define the vector

$$r_u \times r_v = (A, B, C); \qquad A = \frac{\partial(y, z)}{\partial(u, v)}, \qquad B = \frac{\partial(z, x)}{\partial(u, v)}, \qquad C = \frac{\partial(x, v)}{\partial(u, v)}$$

provided that r_u, r_v are not dependent, for then A, B, C do not all vanish and

$$\sigma = [A^2 + B^2 + C^2]^{1/2} = |r_u \times r_v| > 0.$$

We suppose that these conditions hold for almost all u, v, so that $n = [A/\sigma, B/\sigma, C/\sigma]$ is the **unit normal vector,** and the plane perpendicular to it and defined by the vectors r_u and r_v is the tangent plane.

We now define the surface area $S(\mathrm{D})$ of D by

$$S(\mathrm{D}) = \iint\limits_{\Delta} \sigma(u, v)\, du\, dv$$

whenever the integral exists in the Lebesgue sense. Also if $f(r) = f(x, y, z)$ is the value at r of a scalar function f, and similarly for a vector function $f = (p, q, w)$, we say that f is integrable over S and decline surface integrals over D by

$$\iint\limits_{\mathrm{D}} f\, \mathrm{d}S = \iint\limits_{\Delta} f[x(u, v), y(u, v), z(u, v)]\sigma(u, v)\, du\, dv$$

$$\iint\limits_{\mathrm{D}} f \cdot \mathrm{d}S = \iint\limits_{\Delta} f \cdot n\sigma(u, v)\, du\, dv = \iint\limits_{\Delta} f \cdot n\, \mathrm{d}S,$$

respectively whenever these integrals exist in the Lebesgue sense. It is also convenient to define

$$\iint\limits_{\mathrm{D}} f\, \mathrm{d}y\, \mathrm{d}z = \iint\limits_{\Delta} fA\, du\, dv,$$

and so on, so that

$$\iint\limits_{\mathrm{D}} f \cdot \mathrm{d}S = \iint\limits_{\mathrm{D}} p\, \mathrm{d}y\, \mathrm{d}z + q\, \mathrm{d}z\, \mathrm{d}x + w\, \mathrm{d}x\, \mathrm{d}y.$$

It is important to remember that $\iint p\,\mathrm{d}y\,\mathrm{d}z$ is defined only as a Stieltjes integral with respect to u and v, and not an ordinary integral over a set in the (y, z) plane unless the mapping $(u, v)\rightarrow(y, z)$ is one-to-one, and this need not always be the case.

The integral $\iint_X f\,\mathrm{d}S$ of f over a subset X of D which is the image of a measurable subset of Δ is defined as $\iint f\chi_X\,\mathrm{d}S$, where χ_X is the characteristic function of X. It is plainly an absolutely continuous function of X over the completely additive system of subsets of D which are measurable in this sense. As an example of the use of the formulae, we consider the hemisphere defined by parametric equations

$$x=u, \qquad y=v, \qquad z^2=1-u^2-v^2 \quad \text{for} \quad u^2+v^2\leqslant 1,$$

so that

$$r_u=(1, 0, -u/z), \qquad r_v=(0, 1, -v/z),$$
$$A=u/z, \qquad B=v/z, \qquad C=1, \qquad \sigma=1/z.$$

The area of the hemisphere is then

$$\iint_{\Delta} z^{-1}\,\mathrm{d}u\,\mathrm{d}v=2\pi,$$

by Theorem 20 of Chapter 3 with polar coordinates substituted for the cartesian u and v. It is interesting to note how the use of Lebesgue integration takes away any complication related to the fact that z^{-1} is unbounded.

We are now equipped to state and prove the three famous and important theorems on surfaces and their boundary curves, and volumes and their boundary surfaces, which are known under the names of Green, Stokes, and Gauss.

Theorem 2 (Green). *Suppose that the plane domain* D *and its boundary curve* C *are the images of the unit disk* Δ *and its boundary* Γ *under the function* r *defined by* $(u, v)\rightarrow[x(u, v), y(u, v)]=r(u, v)$ *and that* r, r_u, r_v *are continuously differentiable within* Δ *and piecewise continuously differentiable on* C. *Suppose also that* $F=[P(x, y), Q(x, y)]$ *and that* P, Q *are differentiable a.e. and have bounded derivatives. Then* F *is integrable along* C *and if* n *is the unit normal vector to* C,

$$\int_C F\cdot\mathrm{d}s=\int_C P\,\mathrm{d}x+Q\,\mathrm{d}y=\iint_D\left(\frac{\partial Q}{\partial x}-\frac{\partial P}{\partial y}\right)\mathrm{d}x\,\mathrm{d}y$$

$$\int_C F\cdot n\,\mathrm{d}s=\iint_D \operatorname{div} F\,\mathrm{d}x\,\mathrm{d}y.$$

Theorem 3 (Stokes). *Suppose that the surface* D *and its boundary curve* C *are the images in* $\mathcal{R}^3$ *of the unit disk* Δ *and its boundary* Γ *under the function* r *and that* r, r_u, r_v *are continuously differentiable in* D *and piecewise continuously differentiable on* Γ *or, more generally, that* D *can be divided into a finite number of such surfaces. Suppose also that* $F = [P(x, y, z), Q(x, y, z), W(x, y, z)]$ *and that* P, Q, W *are differentiable almost everywhere and have bounded derivatives. Then* F *is integrable along* C *and*

$$\int_C F \cdot ds = \iint_D \operatorname{curl} F \cdot ds = \iint_D \left(\frac{\partial W}{\partial y} - \frac{\partial Q}{\partial z}\right) dy\, dz + \left(\frac{\partial P}{\partial z} - \frac{\partial W}{\partial x}\right) dz\, dx$$

$$+ \left(\frac{\partial Q}{\partial x} - \frac{\partial P}{\partial y}\right) dx\, dy$$

It is obvious that the first part of Theorem 2 is the special case $z = 0$ of Theorem 3 and that the second part of Theorem 2 then follows by replacing (P, Q) by $(-Q, P)$.

We note also that the special case of Green's theorem in which C is itself the unit circle follows from Theorem 15 of Chapter 3 (itself a special case of Green's theorem). We deduce Theorem 3, and hence Theorem 2, from this; and it is sufficient to consider only the case $Q = W = 0$, so that the required conclusion is

$$\int_C P\, dx = \iint_\Delta \left(B\frac{\partial P}{\partial z} - C\frac{\partial P}{\partial y}\right) du\, dv.$$

But

$$\int_C P\, dx = \int_\Gamma P\left(\frac{\partial x}{\partial u} du + \frac{\partial x}{\partial v} dv\right)$$

and it then follows from Green's theorem for the disk, with

$$\left(P\frac{\partial x}{\partial u}, P\frac{\partial x}{\partial v}\right)$$

in place of F, that

$$\int_C P\, dx = \iint \left[\frac{\partial}{\partial u}\left(P\frac{\partial x}{\partial v}\right) - \frac{\partial}{\partial v}\left(P\frac{\partial x}{\partial u}\right)\right] du\, dv$$

$$= \iint_\Delta \left[\frac{\partial P}{\partial u}\frac{\partial x}{\partial v} - \frac{\partial P}{\partial v}\frac{\partial x}{\partial u}\right] du\, dv$$

since the continuity of r_{uu}, r_{vv}, r_{uv} is sufficient to ensure that $x_{uv} = x_{vu}$. The last integrand is

$$\frac{\partial x}{\partial v}\left(\frac{\partial P}{\partial x}\frac{\partial x}{\partial u}+\frac{\partial P}{\partial y}\frac{\partial y}{\partial u}+\frac{\partial P}{\partial z}\frac{\partial z}{\partial u}\right)-\frac{\partial x}{\partial u}\left(\frac{\partial P}{\partial x}\frac{\partial x}{\partial v}+\frac{\partial P}{\partial y}\frac{\partial y}{\partial v}+\frac{\partial P}{\partial z}\frac{\partial z}{\partial v}\right)=B\frac{\partial P}{\partial z}-C\frac{\partial P}{\partial y},$$

as required.

The last theorem of this section relates an integral over the surface of a closed surface to an integral over the region of $\mathcal{R}^2$ which it encloses and can be regarded as an extension of Green's theorem to higher dimensions.

Theorem 4 (Gauss). *Suppose that the closed surface* D *and the region* L *enclosed by it are the images of the surface* Δ *and the interior* Λ *of the unit sphere under the function r defined by*

$$r=[x(\xi,\eta,\zeta),\ y(\zeta,\eta,\zeta),\ z(\xi,\eta,\zeta)]$$

for (ξ,η,ζ) *in* $\Lambda\cup\Delta$. *We suppose that* r, r_u, r_v *are continuously differentiable in* Λ *and piecewise continuously differentiable on* Δ *and that points* (ξ,η,ζ) *on* Δ *are parametrized by*

$$\xi=\xi(u,v),\qquad \eta=\eta(u,v),\qquad \zeta=\zeta(u,v)$$

in such a way that ξ, η, ζ *and their derivatives are continuously differentiable. We suppose also that* $F=(P,R,W)$ *is differentiable with respect to x, y, z with bounded derivatives almost everywhere.*

Then F is integrable over D *and*

$$\iint_{\mathrm{D}} F\cdot \mathrm{d}S=\iiint_{\mathrm{L}}\operatorname{div}F\,\mathrm{d}x\,\mathrm{d}y\,\mathrm{d}z.$$

It is sufficient to consider the case $Q=W=0$, so that we have to prove only that

$$\iint_{\mathrm{D}} P\,\mathrm{d}y\,\mathrm{d}z=\iiint_{\mathrm{L}}\frac{\partial P}{\partial x}\,\mathrm{d}x\,\mathrm{d}y\,\mathrm{d}z. \tag{1}$$

By definition,

$$\iint_{\mathrm{D}} P\,\mathrm{d}y\,\mathrm{d}z=\iint_{\Delta} P\frac{\partial(y,z)}{\partial(u,v)}\,\mathrm{d}u\,\mathrm{d}v$$

$$=\iint_{\Delta} P\left[\frac{\partial(y,z)}{\partial(\eta,\zeta)}\frac{\partial(\eta,\zeta)}{\partial(u,v)}+\frac{\partial(y,z)}{\partial(\zeta,\xi)}\frac{\partial(\zeta,\xi)}{\partial(u,v)}+\frac{\partial(y,z)}{\partial(\xi,\eta)}\frac{\partial(\xi,\eta)}{\partial(u,v)}\right]\mathrm{d}u\,\mathrm{d}v$$

$$=\iint_{\Delta} P\frac{\partial(y,z)}{\partial(\eta,\zeta)}\,\mathrm{d}\eta\,\mathrm{d}\zeta+P\frac{\partial(y,z)}{\partial(\zeta,\xi)}\,\mathrm{d}\zeta\,\mathrm{d}\xi+P\frac{\partial(y,z)}{\partial(\xi,\eta)}\,\mathrm{d}\xi\,\mathrm{d}\eta$$

by partial differentiation.

We can now deduce from the special case of the theorem for a sphere, a consequence of Theorem 16 of Chapter 3, that

$$\iint_D P\,dy\,dz = \iiint_\Lambda \left[\frac{\partial P}{\partial \xi}\frac{\partial(y, z)}{\partial(\eta, \zeta)} + \frac{\partial P}{\partial \eta}\frac{\partial(y, z)}{\partial(\zeta, \xi)} + \frac{\partial P}{\partial \zeta}\frac{\partial(y, z)}{\partial(\xi, \eta)}\right] d\xi\,d\eta\,d\zeta \qquad (2)$$

since the other terms in the integrand are

$$P\left(\frac{\partial}{\partial \xi}\frac{\partial(y, z)}{\partial(\eta, \zeta)} + \frac{\partial}{\partial \eta}\frac{\partial(y, z)}{\partial(\zeta, \xi)} + \frac{\partial}{\partial \zeta}\frac{\partial(y, z)}{\partial(\xi, \eta)}\right)$$

which reduces to zero on expansion. Since the integrand on the right of (2) is

$$\frac{\partial(P, y, z)}{\partial(\xi, \eta, \zeta)} = \frac{\partial(P, y, z)}{\partial(x, y, z)}\frac{\partial(x, y, z)}{\partial(\xi, \eta, \zeta)} = \frac{\partial P}{\partial x}\frac{\partial(x, y, z)}{\partial(\xi, \eta, \zeta)},$$

we have the required conclusion (1) by appealing to Theorem 20 of Chapter 3.

These three theorems have innumerable application in potential theory, fluid mechanics and other branches of physics. In Stokes' theorem, for example, if F is the velocity of a fluid, curl F is the local angular velocity, while $\int_C F \cdot ds$ is the circulation of the flow around the boundary curve. The velocity field is irrotational if curl $F = 0$ everywhere and this is equivalent to the condition that circulation vanishes around any closed curve. Similarly, in Gauss the integral $\iint_D F \cdot dS$ defines the flux or the total outward flow across the boundary of the region L. A vector field in which the total flux across every closed surface vanishes is called solenoidal, and the necessary and sufficient condition for this is that div $F = 0$, and this is also the condition in a field without sources or sinks for the fluid to be incompressible.

4.4 Ergodic theorems and mixing

The material of this section is not strictly geometrical since it relates only to the measure properties of a space without reference to any geometrical or topological properties. But its history and applicability to euclidean spaces suggest that it is not out of place in this chapter. It is not possible to do more than introduce some basic ideas and show how they are expressed in the famous theorems of von Neumann and Birkhoff which gave impetus to the modern theory which is now very extensive. These theorems are treated in the classical idiom in which they were first presented, but only minor changes in notation are needed to show them as part of the theory of unitary operators in Hilbert space which is the natural context in which the more recent development of ergodic theory has taken place.

The starting point is a space $\mathscr{X}$ with a measure μ and its σ-ring of measurable sets. We then suppose that T is a measure-preserving transformation of $\mathscr{X}$ into itself and for any function f defined on $\mathscr{X}$ we define the sequence of functions ϕ_n

$$\phi_n = \frac{1}{n}\sum_{k=0}^{n-1} f_k, \quad f_k(x) = f(T^k x). \tag{1}$$

Ergodic theory is then concerned with the behaviour of ϕ_n as $n \to \infty$, and there is a parallel theory in which the discrete transformations T^k for integral k are replaced by a group of transformations T^α for real α and with the group properties that $T^{\alpha+\beta} = T^\alpha T^\beta$, $T^0 = 1$. The mean ϕ_n is then replaced by $\phi_A = A^{-1}\int_0^A f_\alpha \, d\alpha$, $f_\alpha(x) = f(T^\alpha x)$, and the transformation describes a 'continuous flow' of a hypothetical fluid in the space $\mathscr{X}$. If $\mathscr{X}$ is a euclidean space with μ representing volume, the measure preserving property of T^α means that the fluid is incompressible. We deal here only with the discrete case, but the continuous case is only marginally more difficult and similar proofs apply. It is often helpful in both cases to think of k and α as time variables although this is not relevant to the mathematics.

The connection with operator theory is obvious if we note that the transformation $x \to Tx$ induces a transformation $f \to Uf$, for all real valued functions f in $\mathscr{X}$, defined by $Uf(x) = f(Tx)$,

$$U^k f(x) = f_k(x) = f(T^k x).$$

It is obvious that the operator U is linear and that Uf belongs to $L_p = L_p(\mathscr{X})$ $(p \geq 1)$ whenever f belongs to L_p; and that $\|Uf\|_p = \|f\|_p$. This means that U is an isometry in L_p. In the particularly important case $p = 2$, U has an inverse when T is one-to-one, and U is then a unitary operator in the Hilbert space $L_2(\mathscr{X})$. A large part of the theory can be developed in terms of such a unitary operator in an abstract space, and there is then no need to refer to any underlying measure, although there remains, nevertheless, an important physical interpretation of it. We do not generally assume in what follows that T is one-to-one so that T and U need not have inverses.

The term *ergodic* first appears, both logically and historically, as a property of the transformation T, which is said to be ergodic if $\mathscr{X}$ has no proper subset $\mathscr{X}$ (neither X nor $\mathscr{X} - X$ null) which is invariant under T. In other words, T is ergodic if the condition $T(X) = X$ apart from null sets implies that X is either null or is the whole space $\mathscr{X}$. For example, if $\mathscr{X} = \mathscr{N}$, the transformation $x \to x+1$ is ergodic, while $x \to x+2$ leaves the subsets of even and odd integers both invariant and is therefore not ergodic. The intuitive notion of an ergodic flow is that a particle in a proper subset is not constrained to remain in it, but is forced to wander outside, so that particles become mixed throughout the space.

It is easy to see that T is ergodic if the only invariant functions under the induced operator U are constant, for the characteristic function of a proper subset cannot be constant. Conversely, if T is ergodic and f is not constant, we can choose a real number c so that $\{x;\ f(x) \geqslant c\}$ and $\{x;\ f(x) < c\}$ both have positive measure and the intersection of $\{x;\ f(x) \geqslant c\}$ and $\{x; f(Tx) < c\}$ also has positive measure and f cannot be invariant. The ergodic property can therefore be expressed in terms of U by the characteristic property that U has no invariant function other than constants. The following theorems show how this notion can be made precise. The first is called the *mean ergodic theorem* of von Neumann in which convergence in L_2 is taken as the appropriate description of the asymptotic behaviour of ϕ_n.

Theorem 5 (Mean ergodic theorem of von Neumann). *If T is a measure preserving transformation of a measure space $\mathscr{X}$ and $f \in L_2(\mathscr{X})$ and ϕ_n, f_k are defined as above, then ϕ_n tends in L_2 to a limit f^* as $n \to \infty$. Moreover, f^* is in L_2 and invariant under T.*

Corollary. *If T is ergodic, then f^* is constant and is zero if $\mathscr{X}$ has infinite measure.*

We write $l = \inf \|\sum c_i f_i\|_2$ for all finite sums $\sum c_j f_j$ with $c_j \geqslant 0$, $\sum c_j = 1$ and if $\varepsilon > 0$ choose one particular such sum

$$g = g_0 = \sum_{j=0}^{v} c_j f_j \qquad \text{so that} \qquad \|g\|_2 \leqslant l + \varepsilon.$$

If γ_n is defined as

$$n^{-1} \sum_{k=0}^{n-1} g_k,$$

it then follows from Minkowski's inequality (Theorem 35 of Chapter 2) that $\|\gamma_n\|_2 \leqslant l + \varepsilon$. But

$$g_k = \sum_{j=0}^{v} c_j f_{j+k}, \qquad n\gamma_n = \sum_{k=0}^{n-1} g_k = \sum_{k=0}^{n-1} \sum_{j=0}^{v} c_i f_{j+k}$$

which can be rearranged in the form

$$\sum_{i=0}^{n+v-1} b_i f_i$$

in which it follows from the condition $\sum c_j = 1$ that all but $2v - 1$ of the coefficients b_i are equal to 0. If we compare this sum for $n\gamma_n$ with the corresponding sum

$$\sum_{i=0}^{n-1} f_i \quad \text{for} \quad n\phi_n,$$

we see that

$$n\,\|\phi_n - \gamma_n\|_2 \leq (2v-1)\,\|f\|_2$$

and it follows that

$$l \leq \|\phi_n\|_2 \leq n^{-1}(2v-1)\,\|f\|_2 + l + \varepsilon$$

and, since this holds for all positive ε, we have $\|\phi_n\|_2 \to l$.

Now if n, m are any positive integers, we use the fact that $\phi_n + \phi_m$ has the form $\sum c_i' f_i$ with $c_i' \geq 0$, $\sum c_2' = 2$ to deduce that

$$\|\phi_n + \phi_m\|_2^2 \geq 4l^2$$

and so (after the elementary formula $|z+w|^2 + |z-w|^2 = 2\,|z|^2 + 2\,|w|^2$ for complex values z, w)

$$\|\phi_n - \phi_m\|_2^2 \leq 2\,\|\phi_n\|_2^2 + 2\,\|\phi_m\|_2^2 - 4l^2.$$

Hence $\|\phi_n - \phi_m\| \to 0$ as n, $m \to \infty$ and it follows from Theorem 37 of Chapter 2 that there is a unique function f^* of L_2 for which $\|\phi_n - f^*\|_2 \to 0$. The invariance of f^* follows immediately from the fact that $\phi_n(Tx) \to f^*(Tx)$, while also $\phi_n(Tx) = \phi_n(x) + n^{-1}[f_{n+1}(x) - f_1(x)] \to f^*(x)$ in the L_2 sense and so $f^*(Tx) = f^*(x)$.

The next theorem, often called the *individual ergodic theorem*, is concerned with the pointwise convergence of the same sequences ϕ_n, but is a good deal deeper. It is important to note that the common terminology in which both results are called *ergodic* theorems is slightly illogical and misleading in that ergodicity is a property of a transformation and that it is not assumed of T in the main conclusion of either theorem, but only in special cases arising as corollaries.

Theorem 6 (G. D. Birkhoff). *If T is a measure-preserving transformation on a measure space $\mathscr{X}$ and if $f \in L(\mathscr{X})$, then*

$$\phi_n(x) = n^{-1} \sum_{k=0}^{n-1} f(T^k x)$$

converges a.e. to a function f^* which is integrable $\|f^*\| \leq \|f\|$ and invariant under T. If $\mu(\mathscr{X}) < \infty$, then $\int f^* = \int f$.

Corollary 1. *If $\mu(\mathscr{X}) < \infty$, the sequence ϕ_n converges to f^* in L_1; that is, $\|\phi_n - f^*\| \to 0$.*

Corollary 2. *If T is ergodic, then f^* is constant and is zero if $\mu(\mathscr{X})=\infty$.*

$$\textit{If } \mu(\mathscr{X})<\infty, \textit{ then } f^*(x)=[\mu(\mathscr{X})]^{-1}\int f \quad \textit{a.e.}$$

which can be interpreted as saying that the time mean $f^*(x)$ of $f(T^k x)$ is equal, for almost all x, to its constant space (or phase) mean $[\mu(\mathscr{X})]^{-1}\int f$.

Lemma 1. (*The maximal ergodic theorem*) *if, for almost all x in $\mathscr{X}$, $\phi_n(x)\geqslant 0$ for some $n\geqslant 1$, then $\int f\geqslant 0$.*

We define X_n to be the set in which $\phi_n(x)\geqslant 0$, $\phi_v(x)<0$ for $1\leqslant v\leqslant n-1$, so that X_n are disjoint and their union is X. For x in X_n,

$$(n-v)\phi_{n-v}(T^v x)=\sum_{k=0}^{n-v-1} f_k(T^v x)=\sum_{k=v}^{n-1} f_k(x)=n\phi_n(x)-v\phi_v(x)\geqslant 0,$$

so that $\phi_{n-v}(T^v x)\geqslant 0$ and $T^v x\in X_p$ for some $p\leqslant n-v$. This shows that $T^v X_n$ lies in the union of X_p for $p\leqslant n-v$ and that the sets X_n, $TX_n,\ldots,T^{n-1}X_n$ are disjoint. However, if N is any positive integer, the set $\mathscr{X}_N=\bigcup_{n=1}^{N} X_n$ can be expressed as the union of disjoint sets

$$Q_1=\bigcup_{v=0}^{N-1} T^v X_N,$$

$$Q_2=\bigcup_{v=0}^{N-2} T^v(X_{N-1}-TX_N),$$

$$Q_3=\bigcup_{v=0}^{N-3} T^v(X_{N-2}-TX_{N-1}-T^2X_N),\ldots,$$

in each of which the separate terms are also disjoint. For Q_1, we have

$$\int_{Q_1} f=\sum_{v=0}^{N-1}\int_{X_N} f(T^v X)=N\int_{X_N}\phi_N\geqslant 0,$$

and on repeating the argument for $Q_2, Q_3,\ldots$, we see that $\int_{\mathscr{X}_N} f\geqslant 0$ and the conclusion follows since $f\in L$ and $\mathscr{X}_N\to\mathscr{X}$.

Lemma 2. *If X is an invariant subset, $\beta>0$ and $\sup\phi_n(x)\geqslant\beta$, then $\mu(X)<\infty$ and $\int_X f\geqslant\beta\mu(x)$.*

If G is any measurable subset of X with finite measure, and g is its characteristic function and $\varepsilon>0$, then $h=f-(\beta-\varepsilon)g$ satisfies the conditions of Lemma 1 with X in place of $\mathscr{X}$ and $\int_X f-(\beta-\varepsilon)\mu(G)=\int_X h\geqslant 0$ and the conclusion follows from the fact that X is the limit of an increasing sequence of sets G of finite measure.

Returning now to the proof of the main theorem, we suppose that α, β

are real numbers with $\beta>\alpha$ and let Q be the set of points in which $\limsup \phi_n(X)>\beta>\alpha>\liminf \phi_n(x)$ as $n\to\infty$. It is plain that Q is invariant and it then follows from Lemma 2 applied to f and $-f$ that $\mu(Q)<\infty$ (since either $\beta>0$ or $-\alpha>0$) and that $\beta\mu(Q)\leqslant\int_Q f\leqslant\alpha\mu(Q)$ and therefore $\mu(Q)=0$. Since this holds for all real α and β, it implies that $\phi_n(x)$ converges to a limit $f^*(x)$ a.e. Moreover, since $|f^*|\leqslant\beta$ except in a set of measure $\beta^{-1}\|f\|$, which is arbitrarily small by choice of β, f^* is finite a.e. It is sufficient, since f^+, f^- can be treated separately, to prove the rest of the theorem when $f\geqslant 0$. Then if n and v are integers and $n\geqslant 1$, it follows from Lemma 2 that the sets E_{nv} in which $v\leqslant nf^*(x)<v+1$ have finite measures μ_{nv} and that

$$vn^{-1}\mu_{nv}\leqslant\int_{E_{nv}} f\leqslant(v+1)n^{-1}\mu_{nv},\quad \int_{E_{nv}} f-n^{-1}\mu_{nv}\leqslant\int_{E_{nv}} f^*\leqslant(v+1)n^{-1}\mu_{nv}$$

$$\leqslant(v+1)n^{-1}\int_{E_{nv}} f.$$

If $\varepsilon>0$, we deduce from the last inequality that

$$\int_{f^*\geqslant\varepsilon} f^*\leqslant\sum_{v\geqslant\varepsilon n}\int_{E_{nv}} f^*\leqslant\sum_{v\geqslant\varepsilon n}[1+(\varepsilon n)^{-1}]\int_{E_{nv}} f\leqslant[1+(\varepsilon n)^{-1}]\int f,$$

and this implies that $f^*\in L$ and $\int f^*\leqslant\int f$. Finally, if $\mu(\mathscr{X})<\infty$,

$$\left|\int f^*-\int f\right|\leqslant\sum_{v=0}^{\infty}\left|\int_{E_{nv}} f^*-\int_{E_{nv}} f\right|\leqslant n^{-1}\sum_{v=0}^{\infty}\mu_{nv}\leqslant n^{-1}\mu(\mathscr{X})$$

and we deduce that $\int f^*=\int f$ by letting $n\to\infty$.

To prove Corollary 1, we suppose that $\varepsilon>0$ and define a bounded function g and its associated sequence γ_n so that $\|f-g\|<\varepsilon$. Then

$$\|\phi_n-\gamma_n\|=\|f-g\|<\varepsilon \qquad\text{and}\qquad \|f^*-g^*\|\leqslant\|f-g\|<\varepsilon$$

by the theorem, so that

$$\|\phi_n-f^*\|\leqslant\|\phi_n-\gamma_n\|+\|f^*-g^*\|+\|\gamma_n-g^*\|\leqslant 2\varepsilon+o(1)$$

by Theorem 18 of Chapter 2, and this is sufficient. Corollary 2 follows from the fact that f^* is invariant and therefore constant when T is ergodic. Since f^* is also integrable, it must be zero if $\mu(X)=\infty$ and satisfy $f^*(x)\mu(\mathscr{X})=\int f$ if $\mu(\mathscr{X})<\infty$.

The characteristic property of an ergodic transformation regarded as a flow in a space $\mathscr{X}$ of finite measure is that a particle originally at point x moves successively to the points Tx, $T^2x,\ldots$, and then over a long time interval n lies within a given set X for a proportion $\mu(X)/\mu(\mathscr{X})$ of this time. This is obviously related to the familiar and intuitive idea of the particles becoming mixed by the flow and it is easy to make this idea precise if we suppose that T is one-to-one, and therefore invertible, by considering the behaviour of $\mu(T^{-n}X \cap Y)$ for given measurable sets. If we suppose that $\mu(\mathscr{X})=1$, this is the proportion of the set Y which consists of points which have been transferred there from X after time n, and T is a mixing transformation if it approximates in some sense to $\mu(X)\mu(Y)$ as $n \to \infty$. The extreme case is that in which the limit has the sense of ordinary convergence and $\lim \mu(T^{-n}X \cap Y) = \mu(X)\mu(Y)$ for all measurable X and Y as $n \to \infty$, and T is then said to be *strongly mixing*. The property can be expressed in the functional form that $(U^n f, g) \to 0$ as $n \to \infty$ for all f, g of $L_2(\mathscr{X})$, where U is the unitary operator in L_2 induced by T. Under less stringent conditions, T is called *weakly mixing* if

$$n^{-1} \sum_{k=1}^{n} |\mu(T^{-n}X \cap Y) - \mu(X)\mu(Y)| \to 0 \quad \text{as} \quad n \to \infty$$

for all measurable sets X, Y, and this again has the equivalent functional form that

$$n^{-1} \sum_{k=1}^{n-1} |(U^k f, g) - (f, 1)(g, 1)| \to 0 \quad \text{as} \quad n \to \infty$$

for all f, g, of L_2. It is obvious that a strongly mixing transformation is also weakly mixing. A satisfactory treatment of mixing transformations, including the establishment of necessary and sufficient conditions for a transformation to be either strongly or weakly mixing depends on the theory of the spectral properties of the operator U and is beyond our scope, but the following simple result shows a relationship in one direction and follows easily from what has already been done. Its converse, however, is false.

Theorem 7. *If an invertible measure preserving transformation T on a space of finite measure is weakly (or strongly) mixing, then it is ergodic.*

We deduce from Theorem 4 that $\phi_n \to f^*$ in L_2 and that $\int \phi_n g \to \int f^* g$ for all f, g of L_2. But weak mixing implies that $\int \phi_n g \to \int f \int g$ for all f, g of L_2 so that $\int f^* g = \int f \int g$, and this can be true only if $f^*(x) = \int f$ a.e., and this is the condition for ergodicity.

5
Harmonic analysis

5.1 General forms

Harmonic analysis is the study of relationships of the form

$$\phi(t) \sim \int e^{ixt}\, dF(x) \sim \int e^{ixt} f(x)\, dx \tag{1}$$

which can be interpreted, once a precise meaning has been attached to the non-committal equivalence sign, as the resolution of the function ϕ on $\mathcal{R}$ as an integral (not necessarily in the Lebesgue sense) of pure wave forms e^{ixt} of different frequencies $x/2\pi$ (periods $2\pi/x$). The exponential form is used for convenience, but only trivial modifications are needed to replace it by the trigonometrical form with $\cos xt$ and $\sin xt$.

Harmonic analysis occupies a central position in relation to many branches of mathematics and has innumerable applications in statistics, economics, and science and particularly in the treatment of any wave phenomenon in which the function F defines the spectrum of frequencies of the light, for example, constituting the complex wave form ϕ. Moreover, quite apart from its great formal elegance, it is a field in which the ideas of measure and Lebesgue integration are immediately seen as appropriate and natural. In particular, the Stieltjes form of (1) exploits the power and economy to be gained from considering harmonic analysis in as general a context as possible.

The relationship has to be interpreted in different ways according to the properties of the functions ϕ and F, but there are important formal relationships which apply throughout the whole theory. Either ϕ or F can be regarded as given and taken as the starting point, but this must be followed by a precise account of the definition of the other and the meaning to be attached to formulae (1). It is natural to think of the formula as a mapping or transformation between spaces of function ϕ and F and to look for general properties of the transformation. It is clear, for example, that the transformation will be linear under any reasonable interpretation and we find that in most cases it is one-to-one. Purely heuristic manipulation suggests the plausibility of the *inversion formulae*

$$F(x) - F(a) \sim \int \frac{e^{-ixt} - e^{-iat}}{-2\pi i t}\, \phi(t)\, dt, \qquad f(x) = F'(x) \sim \frac{1}{2\pi} \int e^{-ixt} \phi(t)\, dt$$

and the *convolution formulae*

$$L(x)=\int F(x-y)\,\mathrm{d}G(y)\sim\frac{1}{2\pi}\int \mathrm{e}^{-\mathrm{i}xt}\phi(t)\gamma(t)\,\mathrm{d}t,\qquad \phi(t)\gamma(t)\sim\int \mathrm{e}^{\mathrm{i}xt}\,\mathrm{d}L(x)$$

when

$$\phi(t)\sim\int \mathrm{e}^{\mathrm{i}xt}\,\mathrm{d}F(x),\qquad \gamma(t)\sim\int \mathrm{e}^{\mathrm{i}xt}\,\mathrm{d}G(x).$$

These and other formulae are established on a proper basis by dealing separately with the most important and familiar cases.

5.2 Fourier series

The earliest and best known case of harmonic representation is the Fourier series of a periodic function ϕ on $\mathscr{R}$ with period 2π and integrable over the interval $-\pi\leqslant x<\pi$ (and so integrable over every finite interval). The Fourier coefficient c_n is defined for every integer n by

$$c_n=\frac{1}{2\pi}\int_{-\pi}^{\pi}\mathrm{e}^{-\mathrm{i}nt}\phi(t)\,\mathrm{d}t.$$

Since

$$\int_{-\pi+a}^{\pi+a}=\int_{-\pi}^{\pi}+\int_{\pi}^{\pi+a}-\int_{-\pi}^{-\pi+a}$$

it follows from the periodicity of $\mathrm{e}^{-\mathrm{i}nt}\phi(t)$ that the interval $(-\pi,\pi)$ may be replaced by any other interval of length 2π. We write

$$\phi(t)\sim\sum_{-\infty}^{\infty}c_n\mathrm{e}^{\mathrm{i}nt}$$

and call the series the Fourier series of ϕ. The signs $\pm\infty$ can usually be omitted without ambiguity. The series is clearly of the form (1) with F defined by

$$F(x)-F(a)=\sum_{a<n\leqslant x}c_n,$$

but the meaning of the equivalence sign is still open.

The most natural meaning and the one to be expected in practice is that the series converges to a sum equal to $\phi(x)$. The search for conditions for this to hold has been a dominant part of the theory and there is an extensive literature on it. A salient feature of the theory is that there is no final answer in terms of necessary *and* sufficient conditions on ϕ for the convergence of its Fourier series at a specified point. We have rather a catalogue of sufficient conditions and we confine our consideration of these to a few familiar cases. They all involve the behaviour of ϕ in a neighbourhood of a point t and require the existence of the limiting values $\phi(t+0)$, $\phi(t-0)$ of $\phi(u)$ as u tends to t for the right or left, respectively. In fact, it is sufficient for this to hold only for the values of $\phi(u)$ outside a null set, since the coefficients c_n are not affected by the values of ϕ in such a set. Before dealing with convergence, we need two general theorems which are widely applicable.

Theorem 1 (Riemann–Lebesgue). *If $\phi \in L(-\infty, \infty)$, then $\int e^{ixt}\phi(t)\,dt \to 0$ as $x \to \pm\infty$. In particular, if c_n are Fourier coefficients of an integrable periodic function, then $c_n \to 0$ as $n \to \pm\infty$.*

Since $e^{ixt} = -e^{i(xt+\pi)} = e^{ix(t+\pi/x)}$, we have

$$\begin{aligned}\left|2\int e^{ixt}\phi(t)\,dt\right| &= \left|\int e^{ixt}\phi(t)\,dt - \int e^{ix(t+\pi/x)}\phi(t)\,dt\right| \\ &= \left|\int e^{ixt}[\phi(t)-\phi(t-\pi/x)]\,dt\right| \leqslant \int |\phi(t)-\phi(t-\pi/x)|\,dt \\ &= o(1) \text{ as } x \to \pm\infty \qquad \text{by Theorem 10(ii) of Chapter 3.}\end{aligned}$$

For the corollary we consider the (non-periodic) function with values equal to those of ϕ for $-\pi \leqslant x < \pi$ and zero elsewhere.

Theorem 2 (Localization). *If $\phi \in L(-\pi, \pi)$, $\phi(t) \sim \sum c_n e^{int}$ and $\delta > 0$, then*

(i)

$$s_n(t) = \sum_{v=-n}^{n} c_v e^{ivt} = \int_{-\pi}^{\pi} \phi(t-u) D_n(u)\,du,$$

where

$$D_n(u) = \frac{1}{2\pi}\sum_{v=-n}^{n} e^{ivu} = \frac{\sin(n+\frac{1}{2})u}{2\pi \sin(u/2)}, \qquad \int_{-\pi}^{\pi} D_n(u)\,du = 1.$$

(ii)

$$s_n(t) = \frac{1}{\pi}\int_{-\delta}^{\delta} \phi(t-u)\frac{\sin nu}{u}\,du + o(1),$$

and the convergence or divergence of $s_n(t)$ does not depend on values of ϕ outside $(-\delta, \delta)$.

Part (i) follows almost trivially from the definition of c_n, since

$$\begin{aligned} s_n(t) &= \frac{1}{2\pi}\sum_{-n}^{n} e^{ivt}\int_{-\pi}^{\pi} e^{-ivu}\phi(u)\,du \\ &= \int_{-\pi}^{\pi} \phi(u)D_n(t-u)\,du = \int_{-\pi}^{\pi} \phi(t-u)D_n(u)\,du\end{aligned}$$

and we note, in integrating the series for $D_n(u)$, that $\int_{-\pi}^{\pi} e^{ivu} = 0$ when $v \neq 0$. For part (ii), we note first that

$$\int_{-\pi}^{t-\delta} + \int_{t+\delta}^{\pi} \phi(t-u)D_n(u)\,du = o(1) \quad \text{as} \quad n \to \infty$$

by Theorem 1, since $\phi(t-u)(\sin u/2)^{-1}$ is integrable in each range of integration. Then

$$s_n(t) - \frac{1}{\pi}\int_{-\delta}^{\delta} \phi(t-u)\frac{\sin nu}{u}\,du = \frac{1}{2\pi}\int_{-\delta}^{\delta} \phi(t-u)\left(\cot\frac{u}{2} - \frac{2}{u}\right)\sin nu\,du$$
$$+\frac{1}{2\pi}\int_{-\delta}^{\delta} \phi(t-u)\cos nu\,du + o(1) = o(1)$$

by Theorem 1, since $\cot(u/2) - 2/u$ is bounded and $\phi(x-y)$ is integrable in $(-\delta, \delta)$, and this completes the proof.

Theorem 3 (Convergence at a point). *If ϕ is periodic 2π and integrable and if $\phi(t \pm 0)$ both exist, then the Fourier series of ϕ converges to $\frac{1}{2}[\phi(t+0) + \phi(t-0)]$ provided that any one of the following conditions is satisfied.*

(i) *$\phi(t \pm u) - \phi(t \pm 0) = O(u^{\alpha})$ for some positive constant α, as $u \to +0$.*
(ii) *ϕ is differentiable on the left and on the right at t in the sense that both $h^{-1}[\phi(t+h) - \phi(t+0)]$ and $h^{-1}[\phi(t-h) - \phi(t-0)]$ tends to limits (which need not be the same) as $h \to 0$.*
(iii) *ϕ is differentiable at x.*
(iv) *ϕ is of bounded variation in some interval with centre x.*

We can write

$$s_n(t) - \tfrac{1}{2}[\phi(t+0) + \phi(t-0)] = \frac{1}{\pi}\int_0^{\delta} \psi(u)\frac{\sin nu}{u}\,du + o(1),$$

where $\psi(u) = \phi(t+u) - \phi(t+0) + \phi(t-u) - \phi(t-0)$, with $\psi(+0) = 0$.

The conclusion follows in (i) from Theorem 1, since $\psi(u) = o(u^{\alpha-1})$ near 0 and ψ is therefore integrable, and (ii), (iii) follow from the special case $\alpha = 1$. In (iv), ψ is also of bounded variation in $(0, \delta)$ and we have, by partial integration,

$$\int_0^{\delta} \psi(u)\frac{\sin nu}{u}\,du = \int_0^{\delta} P_n(u)\,d\psi(u), \qquad P_n(u) = \int_u^{\delta} \frac{\sin nv}{v}\,dv,$$

and the conclusion follows from Theorem 18 of Chapter 2 since $P_n(u) \to 0$ for every u as $n \to \infty$ and $P_n(u)$ is bounded for all n, u by the convergence of

$$\int v^{-1} \sin v\,dv.$$

The Fourier coefficients and series of a given function ϕ are defined uniquely and the converse, that a periodic function is defined uniquely almost everywhere by its Fourier series, is also true but not quite so obvious. It is, in fact, a consequence of the last theorem, and the following useful theorem about the integration of Fourier series.

Theorem 4. *If $\phi \in L(-\pi, \pi)$ and $\phi \sim \sum c_n e^{int}$, then*

$$\Phi(t) = \int_0^t \phi(u)\,du = c_0 t + \sum_{n \neq 0} \frac{c_n}{in} e^{int} \quad a.e.$$

In other words, any Fourier series can be integrated term by term.

Since $\Phi(t+2\pi) - \Phi(t) = \int_t^{t+2\pi} \phi(u)\,du = \int_{-\pi}^{\pi} \phi(u)\,du = 2\pi c_0$, it is plain that $\Phi(t) - c_0 t$ is periodic 2π. Then, by partial integration,

$$\frac{1}{2\pi}\int_{-\pi}^{\pi} e^{-int}[\Phi(t) - c_0 t]\,dt = \frac{1}{2\pi in}\int_{-\pi}^{\pi} e^{-int}\phi(t)\,dt = \frac{c_n}{in} \qquad (n \neq 0),$$

$$\Phi(t) - c_0 t \sim \sum_{n \neq 0} \frac{c_n}{in} e^{int},$$

and the conclusion follows from Theorem 3 and the fact that Φ is differentiable a.e. As an immediate corollary, we have

Theorem 5 (Uniqueness). *If integrable periodic functions ϕ, ψ have the same Fourier coefficients, then $\phi(t) = \psi(t)$ a.e.*

It is sufficient to show that $c_n = 0$ for all n implies that $\phi(t) = 0$ a.e. After Theorem 4, $c_n = 0$ implies that

$$\Phi(t) = \int_0^t \phi(u)\,du = 0$$

almost everywhere and hence for all x, since $\Phi(t)$ is continuous, and so $\phi(t) = \Phi'(t) = 0$ a.e.

It is important to note, however, that there is still a basic lack of symmetry in the relationship between a function and its Fourier series. While every integrable function ϕ has a Fourier series and is defined uniquely by it, it is not the case that every trigonometrical series

$$\sum c_n e^{int},$$

even if it converges a.e., is a Fourier series. There are well known conditions ($\sum |c_n| < \infty$, $\sum |c_n|^2 < \infty$ for example) which are sufficient for c_n to be Fourier coefficients, but no general condition which is both necessary and sufficient.

The general theory of infinite series admits limit processes more extensive than ordinary convergence, and these are of importance in the case of Fourier series. We mention only two such methods of summation. The series $\sum a_n$ is said to be summable C (Cesaro) to sum s if $\sigma_n \to s$ as $n \to \infty$ and

$$\sigma_n = \frac{1}{n+1} \sum_{v=0}^{n} s_v;$$

and it is summable A (Abel) if $\sum a_n r^n$ converges to $s(r)$ for $0 \leqslant r < 1$ and $s(r) \to s$ when $r \to 1$. It can be shown without difficulty that both methods are consistent in the sense that they sum any convergent series to its proper sum, and that A is stronger than C in that any C-summable series is also A-summable and with the same sum.

Theorem 6. *The Fourier series of function ϕ of $L(-\pi, \pi)$ is summable* C *(and therefore also* A*) to the value*

$$\tfrac{1}{2}[\phi(t+0) + \phi(t-0)]$$

at any point t at which $\phi(t+0)$, $\phi(t-0)$ both exist and, in particular, at any point of continuity of ϕ.

It is easy to confirm that

$$\sigma_n(t) = \int \phi(t-u) K_n(u)\,du,$$

$$K_n(u) = \frac{1}{2\pi(n+1)} \left(\frac{\sin(n+1)u/2}{\sin u/2}\right)^2,$$

$$\int K_n(u)\,du = 1$$

and that

$$\sigma_n(t) - \tfrac{1}{2}[\phi(t+0) + \phi(t-0)] = \int_0^{\pi} \psi(u) K_n(u)\,du,$$

where $\psi(u) = \phi(t+u) - \phi(t+0) + \phi(t-u) - \phi(t-0)$ is integrable in $(0, \pi)$ and tends to 0 as $u \to +0$. Then if $\varepsilon > 0$ and we define $\delta > 0$ so that $|\psi(u)| \leqslant \varepsilon$ for $0 < u \leqslant \delta$, and use the inequality $\sin u/2 > u/\pi$ for $0 \leqslant u \leqslant \pi$, we have

$$|\sigma_n(t) - \tfrac{1}{2}[\phi(t+0) + \phi(t-0)]|$$

$$\leqslant \varepsilon \int_0^{\delta} K_n(u)\,du + \frac{\pi}{2(n+1)\delta^2} \int_{\delta}^{\pi} |\psi(u)|\,du \leqslant \varepsilon + o(1)$$

as $n \to \infty$, and this is sufficient.

Comparison of this theorem with Theorem 4 reflects the obvious fact that a series is more likely to be summable than convergent and this is confirmed by other results. It can be shown, for example, that a Fourier series is summable, but not necessarily convergent, almost everywhere. The fact that summability is so much easier to establish is due to the crucial difference between the kernels D_n and K_n in that $K_n(u) \geq 0$ and so $\int |K_n(u)|\, du = \int K_n(u)\, du = 1$, while $\int |D_n(u)|\, du \to \infty$ as $n \to \infty$.

We now come to the theory of functions of $L_2(-\pi, \pi)$ and their Fourier series which has widespread and important applications. The theory also has a particularly satisfying symmetry and completeness which is not apparent in many areas of harmonic analysis. We first note the obvious fact that L_2 is a proper subclass of L for a finite interval (but not for an infinite interval).

Theorem 7 (Riesz–Fischer). (i) *If $\sum |c_n|^2 < \infty$ there is a function ϕ of L_2 with Fourier coefficients c_n, and therefore defined uniquely by c_n.*

(ii) *ϕ is the limit in L_2 of the partial sums s_n, so that $\|s_n - \phi\|_2 \to 0$ as $n \to \infty$.*

Lemma. *If $p(t) = \sum b_n e^{int}$ is any trigonometrical polynomial, then*

$$\|p\|_2^2 = 2\pi \sum |b_n|^2.$$

For

$$\|p\|_2^2 = \int_{-\pi}^{\pi} p(t)\overline{p(t)}\, dt = \int_{-\pi}^{\pi} \sum_n \sum_m b_n e^{int} \bar{b}_m e^{-imt}\, dt$$

$$= \sum_n \sum_m b_n \bar{b}_m \int_{-\pi}^{\pi} e^{i(n-m)t}\, dt = 2\pi \sum |b_n|^2$$

since $\int_{-\pi}^{\pi} e^{ivt}\, dt = 0$ for $v \neq 0$.

It follows from the lemma that, for any positive integers n, m $(n \geq m)$,

$$s_n(t) - s_m(t) = \sum_{m+1}^{n} (c_v e^{ivt} + c_{-v} e^{-ivt})$$

$$\|s_n - s_m\|_2^2 = \sum_{m+1}^{n} |c_v|^2 + |c_{-v}|^2 \to 0 \quad \text{as} \quad n, m \to \infty.$$

Then, by Theorem 37 of Chapter 2, there is a unique function ϕ of L_2 such that $\|s_m - \phi\|_2 \to 0$ as $m \to \infty$. If c_n' are the Fourier coefficients of ϕ,

$$c_n' - c_n = \frac{1}{2\pi} \int_{-\pi}^{\pi} e^{-int}[\phi(t) - s_m(t)]\, dt$$

for $m \geq n$, and so

$$|c_n' - c_n| \leq (2\pi)^{-1} \|\phi - s_m\| \leq (2\pi)^{-1/2} \|\phi - s_m\|_2,$$

and this tends to 0 as $m \to \infty$. It follows that $c_n' = c_n$, as required.

Theorem 8 (Parseval). *If* $\phi \in L_2$, $\phi(t) \sim \sum c_n e^{int}$, *then* $\sum |c_n|^2 < \infty$ *and*

$$\|\phi\|_2^2 = \int_{-\pi}^{\pi} |\phi(t)|^2 \, dt = 2\pi \sum |c_n|^2.$$

Using the Lemma of Theorem 7,

$$\begin{aligned}
\|\phi - s_m\|_2^2 &= \int_{-\pi}^{\pi} \left[\phi(t) - \sum_{-m}^{m} c_n e^{int}\right]\left[\bar{\phi}(t) - \sum_{-m}^{m} \bar{c}_\upsilon e^{-i\upsilon t}\right] dt \\
&= \int_{-\pi}^{\pi} |\phi(t)|^2 \, dt - \sum_{-m}^{m} c_n \int_{-\pi}^{\pi} e^{int} \bar{\phi}(t) \, dt \\
&\quad - \sum_{-m}^{m} \bar{c}_\upsilon \int_{-\pi}^{\pi} e^{-i\upsilon t} \phi(t) \, dt + 2\pi \sum_{-m}^{m} |c_n|^2 \\
&= \int_{-\pi}^{\pi} |\phi(t)|^2 \, dt - 2\pi \sum_{-m}^{m} |c_n|^2
\end{aligned}$$

and therefore

$$2\pi \sum_{-m}^{m} |c_n|^2 \leqslant \|\phi\|_2^2,$$

$\sum |c_n|^2$ converges and we have *Bessel's inequality*

$$2\pi \sum |c_n|^2 \leqslant \|\phi\|_2^2.$$

For the inequality in the opposite direction, it follows from Theorem 7(ii) that $\|s_m - \phi\|_2 \to 0$ as $m \to \infty$ and so

$$\begin{aligned}
\|\phi\|_2 = \|s_m + (\phi - s_m)\|_2 &\leqslant \|s_m\|_2 + \|\phi - s_m\|_2 \\
&\leqslant (2\pi \sum |c_n|^2)^{1/2} + \|\phi - s_m\|_2
\end{aligned}$$

and the conclusion follows on letting $m \to \infty$.

As an immediate corollary we have

Theorem 9. *If* $\phi \in L_2$ *and* $\varepsilon > 0$, *we can find a trigonometrical polynomial* p *such that* $\|\phi - p\|_2 < \varepsilon$. *Among polynomials of degree* n, *the closest approximation to* ϕ *in* L_2 *is given when* p *is the partial sum* s_n *of the Fourier series of* ϕ.

5.3 Convolutions

A convolution, in the most general sense, is an integral $\int \phi(x-y) \, d\Gamma(y)$ over a group (with group operation and inverse on elements x, y denoted by $x+y$, $x-y$) in which a measure Γ is defined. The integral is in the usual Lebesgue sense, so that an order of magnitude on ϕ is needed as well as appropriate measurability conditions. These present no problem as the groups we use here are familiar groups of real numbers and all that is necessary is that the functions used should be Borel measurable. In fact, we do not really need the ideas or terminology of group theory except in so far as they provide a helpful insight into the structural differences between different parts of harmonic analysis and the reasons why analogies between systems involving different groups are imperfect and not to be pressed too far.

We write

$$\phi * \mathrm{d}\Gamma(t) = \int \phi(t-u)\,\mathrm{d}\Gamma(u)$$

in the general case and $\phi * \gamma(t) = \int \phi(t-u)\gamma(u)\,\mathrm{d}u$ when Γ is absolutely continuous with Radon–Nikodym derivative γ with respect to some basic measure. It is then plain that the convolution operation, when it is defined, is commutative and associative and distributive with respect to addition of values of the functions. The groups we consider are

(i) the group $\mathscr{R}$ of all real numbers with Lebesque measure and ordinary addition as the operator.
(ii) the circle group $\mathscr{C}$ of rotations—or real numbers $-\pi \leq x \leq \pi$ with Lebesgue measure and equivalence and addition modulo 2π.
(iii) the group $\mathscr{N}$ of integers with ordinary addition and each given unit measure.

Extension to n-dimensional vectors present no difficulty and further generalizations are possible. Although the notation

$$\|c\|_p = \left(\int |c(t)|^p\,\mathrm{d}t\right)^{1/p}$$

still has a valid interpretation when $\mathscr{N}$ is the group, it is more natural to use the common notation

$$\|c\|_p = \left(\sum |c_n|^p\right)^{1/p}$$

for a sequence $c = \{c_n\}$.

Theorem 10. *Suppose that Γ is a measure over one of the groups $\mathscr{R}$ or $\mathscr{C}$ and that $\int |\mathrm{d}\Gamma| < \infty$. Then*

(i) *if $\phi \in L_p$, $p \geq 1$, then $\psi = \phi * \mathrm{d}\Gamma$ exists almost everywhere and belongs to L_p with $\|\psi\|_p \leq \int |\mathrm{d}\Gamma|\,\|\phi\|_p$*
(ii) *if $|\phi(t)| \leq A$, then $|\psi(t)| \leq A \int |\mathrm{d}\Gamma|$*

and if ϕ is uniformly continuous, so is ψ.

The conclusion follows immediately from the definition of ψ and Minkowski's inequality (Theorem 35 of Chapter 2)

Theorem 11. *Suppose that integrals are defined over $\mathscr{R}$ or $\mathscr{C}$ and that $\gamma \in L$. Then*

(i) *if $\phi \in L_p$, $p \geq 1$, then $\psi = \phi * \gamma$ exists almost everywhere and belongs to L_p with $\|\psi\|_p \leq \|\gamma\| \, \|\phi\|_p$.*
(ii) *if $|\phi(x)| \leq A$ a.e. then $|\psi(t)| \leq A \|\gamma\|$ for all t and ψ is uniformly continuous.*

Part (i) is a straight corollary of Theorem 10(i), but part (ii) has different hypotheses from those of Theorem 10(ii) and cannot be derived from it. It follows easily, however, from the facts that

$$|\psi(t)| \leq A \int |\gamma(u)| \, du$$

and

$$|\psi(t+\delta) - \psi(t)| = \left| \int [\gamma(t+\delta-u) - \gamma(t-u)] \phi(u) \, du \right| \leq A \, \|\gamma(t+\delta) - \gamma(t)\|,$$

and the conclusion follows from Theorem 10(ii) of Chapter 3.

Theorem 12. *Suppose that integrals are defined over $\mathscr{R}$ or $\mathscr{C}$ and that $p > 1$, $1/p' = 1 - 1/p$, $\phi \in L_p$, $\gamma \in L_{p'}$. Then $\psi(t) = \phi * \gamma(t)$ exists for all t, $|\psi(t)| \leq \|\phi\|_p \, \|\gamma\|_{p'}$ and ψ is uniformly continuous, and $\psi(t) \to 0$ as $t \to \pm\infty$.*

The first conclusion follows from Hölder's inequality (Theorem 34 of Chapter 3). The uniform continuity of ψ follows for the fact that

$$|\psi(t+\delta) - \psi(t)| \leq \|\phi(t+\delta) - \phi(t)\|_p \, \|\gamma\|_{p'},$$

which is uniformly small with δ by Theorem 10(ii) of Chapter 3.

Finally, if $\varepsilon > 0$, we can define l so that $\phi = \phi_1 + \phi_2$, $\gamma = \gamma_1 + \gamma_2$, $\phi_1(t) = \gamma_1(t) = 0$ for $|t| \geq l$, $\|\phi_2\|_p < \varepsilon$, $\|\gamma_2\|_{p'} < \varepsilon$. Then, for $t \geq 2l$, $\phi_1 * \gamma_1 = 0$ and $|\phi * \gamma| \leq \varepsilon \, \|\gamma_2\|_{p'} + \varepsilon \, \|\phi_2\|_p + \varepsilon^2$, which is sufficient to show that $\psi \to 0$.

These theorems show that the integrability, boundedness, continuity and other properties of 'smoothness' of a convolution correspond to the better behaved of its components and that convoluting any function with a smoother one improves its behaviour accordingly. The best and most useful example is that of the *sliding average* in which $\gamma(t) = (2\delta)^{-1}$ for $|t| \leq \delta$ and $\gamma(t) = 0$ for $|t| > \delta$. Then if ϕ is integrable, $\gamma * \phi$ is integrable and absolutely continuous, $\gamma * \gamma * \phi$ has an absolutely continuous first derivative, and so on. It must be noted, however, that the same does not hold in regard to orders of magnitude for large values of t, for the convolution generally behaves more like the larger of its two components. Thus, if $\phi(t) = (1+t^2)^{-1}$, $\gamma(t) = (1+t^2)^{-2}$, then $\gamma * \phi$ is of order t^{-2} but not t^{-4} for large t.

The next theorem gives the analogies of these results for sequences and the proof is similar but simpler, and not repeated.

Theorem 13.

(i) *If* $\|d\|<\infty$, $|c_n|\leqslant A$, *then* $|h_n|\leqslant A\,\|d\|$ *for all* n.
(ii) *If* $\|d\|_{p'}<\infty$, $\|c\|_p<\infty$, $p\geqslant 1$, *then* $|h_n|\leqslant\|c\|_p\,\|d\|_{p'}$ *and* $h_n\to 0$ *as* $n\to\pm\infty$.
(iii) *If* $\|d\|<\infty$, $\|c\|_p<\infty$, $p\geqslant 1$, *then*

$$\|h\|_p\leqslant\|d\|\,\|c\|_p.$$

In each case, the series $\sum c_{n-v}d_v$ *converges absolutely to* h_n.

The application of the foregoing results on convolution to Fourier series and coefficients in $\mathcal{N}$ of integrable functions on $\mathscr{C}$ is very straightforward and the following results require little in the way of proof.

Theorem 14. *If* ϕ, γ *are periodic* 2π *and integrable, and if*

$$\phi\sim\sum c_n e^{int},\qquad \gamma\sim\sum d_n e^{int},$$

then

$$\psi(t)=\phi*\gamma(t)\sim 2\pi\sum c_n d_n e^{int}.$$

For

$$\begin{aligned}\int e^{-int}\psi(t)\,dt&=\int e^{-int}\int\phi(t-u)\gamma(u)\,du\\&=\int\gamma(u)\,du\int e^{-int}\phi(t-u)\,dt\\&=\int e^{-inu}\gamma(u)\,du\int e^{-int}\phi(t)\,dt=4\pi^2 c_n d_n.\end{aligned}$$

The theorem says nothing about the convergence of the series, but under stronger L_2 conditions on ϕ, γ we can say much more.

Theorem 15. *If the conditions of Theorem* 14 *hold and also* $\phi,\gamma\in L_2$, *then* $\sum|c_n d_n|<\infty$.

This follows from Parseval (Theorem 8) and the Schwartz inequality. (Theorem 34 of Chapter 2)

Theorem 16. *Under the conditions of Theorem* 15,

$$\phi(t)\gamma(t)\sim\sum h_n e^{int},\qquad\textit{where}\qquad h_n=\sum c_{n-v}d_v.$$

Corollary.

$$\sum\bar{c}_n d_n=\frac{1}{2\pi}\int\phi(t)\bar{\gamma}(t)\,dt$$

(*with Parseval's theorem as the special case* $\phi=\gamma$).

The series $\sum c_{n-v}d_v$ converges, by Schwartz, and

$$2\pi \sum_{-m}^{m} c_{n-v}d_v = \sum_{v=-m}^{m} c_{n-v} \int e^{-ivt}\gamma(t)\,dt = \int \gamma(t) \sum_{-m}^{m} c_{n-v} e^{-ivt}\,dt$$

$$= \int e^{-int}\gamma(t) \sum_{-m}^{m} c_{n-v} e^{i(n-v)t}\,dt = \int e^{-int}\gamma(t) q_m(t)\,dt$$

$$= \int e^{-int}\gamma(t)\phi(t) + \int e^{-int}\gamma(t)[q_m(t) - \phi(t)]\,dt$$

where, for each fixed n, $\|\phi - q_m\|_2^2 = 2\pi \sum_{|v-n|>m} |c_m|^2 \to 0$ as $v \to \infty$. The conclusion follows from Schwartz on letting $m \to \infty$.

5.4 Fourier transforms

We consider now the representation (1) in cases in which F is absolutely continuous in every finite interval, and has derivative f a.e., so that the formula becomes

$$\phi(t) \sim \int e^{ixt} f(x)\,dx.$$

The easiest case, and the one most closely analogous to that of Fourier series, arises when $\phi \in L(-\infty, \infty)$ and we define f for all values of y by

$$f(x) = \frac{1}{2\pi} \int e^{-ixt}\phi(t)\,dt \tag{2}$$

and call f the Fourier transform of ϕ. The theorems in this part of the theory are parallel to those for series and hardly any extra proof is needed.

Theorem 17. *If $\phi \in L$, its Fourier transform f is bounded and uniformly continuous and $f(x) \to 0$ as $x \to \pm\infty$.*

This follows from

$$2\pi\, |f(x)| \leq \int |\phi(t)|\,dt,$$

and

$$2\pi\, |f(x+h) - f(x)| = \left| \int e^{ixt}(e^{iht} - 1)\phi(t)\,dt \right|$$

$$\leq \int |e^{iht} - 1|\, \phi(t)\,dt = o(1)$$

uniformly in x as $h \to 0$.

Theorem 18. *If $\phi \in L$ and $\phi(t) \sim \int e^{ixt} f(x)\,dx$, then*

$$\phi_a(t) = \int_{-a}^{a} e^{ixt} f(x)\,dx = \frac{1}{\pi}\int \phi(t-u)\frac{\sin au}{u}\,du$$

and the convergence of $\phi_a(t)$ depends only on the local properties of ϕ near t, and sufficient conditions for this are given by any conditions which ensure that convergence of the Fourier series of a periodic function taking the same values as those of ϕ in a neighbourhood of t.

There is a similar analogy for C-summability in that

$$\sigma_b(t) = \frac{1}{b}\int_0^b \phi_s(t)\,da = \frac{2}{\pi b}\int \phi(t-u)\left(\frac{\sin bu/2}{u}\right)^2 du$$

and the integral $\int e^{ixt} f(x)\,dx$ is summable to $\frac{1}{2}[\phi(t+0)+\phi(t-0)]$ at any point t at which $\phi(t \pm 0)$ exist.

Theorem 19 (Uniqueness). *If functions ϕ, ψ of L have the same Fourier transform, they are equal a.e.*

It is sufficient to show that $\phi \sim 0$ implies that $\phi(t)=0$ a.e. If

$$\Phi(t) = \int_0^t \phi(u)\,du,$$

and $h>0$, it is clear that $\Phi(t+h)-\Phi(t)$ is integrable and $\Phi(t+h)-\Phi(t) \sim 0$. But $\Phi(t+h)-\Phi(t)$ is differentiable a.e. and, by Theorems 3(ii) and 18, $\Phi(t+h)=\Phi(t)$ for all t and h, and hence $\phi(t)=0$ a.e.

Theorem 20. *If $\phi \in L$, $\phi(t) \sim \int e^{ixt} f(x)\,dx$ and $f \in L$, then*

$$\phi(t) = \int e^{ixt} f(x)\,dx \quad a.e.$$

That is,

$$f(x) \sim \frac{1}{2\pi}\int e^{-ixt}\phi(t)\,dt.$$

Since $f \in L$, we have

$$\int e^{ixt} f(x)\frac{\sin hx}{hx}\,dx = \int f(x)\,dx \frac{1}{2h}\int_{t-h}^{t+h} e^{ixu}\,du$$
$$= \frac{1}{2h}\int_{t-h}^{t+h} du \int e^{ixu} f(x)\,dx = \frac{1}{2\pi}\int_{t-h}^{t+h} \phi(u)\,du$$

and it then follows from Theorem 18 of Chapter 2 and Theorem 11 of

Chapter 3, on letting $h \to 0$, that

$$\int e^{ixt} f(x)\,dx = \phi(t) \quad \text{a.e.}$$

as required.

This theorem indicates a certain symmetry between a function and its transform in the very special case in which both belong to L but this should not obscure the fundamental lack of symmetry in the fact that, while every function of L has a Fourier transform, there is no necessary and sufficient conditions on a function which ensures that it is such a transform. To arrive at a more satisfactory kind of symmetry, we have to consider functions in L_2.

The first thing to notice in proceeding from L to L_2 over an *infinite* range is that neither space is included in the other in the way that L_2 is a proper subspace of L for a finite interval. Obvious examples are $(1+|x|)^{-1}$ in L_2 but not L and $(1+|x|)^{-1}x^{-1/2}$ in L but not L_2. This means we cannot define the transform of a function ϕ of L_2 by the formula (2), since $e^{-ixt}\phi(t)$ need not be in L, and we have to proceed less directly. The first step is a theorem about functions which are in both L and L_2.

Theorem 21. *If $\phi \in L \cap L_2$ and*

$$\phi(t) \sim \int e^{ixt} f(x)\,dx$$

(so that $f(x) = (1/2\pi)\int e^{-ixt}\phi(t)\,dt$), and if

$$f_a(x) = \frac{1}{2\pi}\int_{-a}^{a} e^{-ixt}\phi(t)\,dt,$$

then $f \in L_2$, $2\pi\,\|f\|_2^2 = \|\phi\|_2^2$ and $\|f_a - f\|_2 \to 0$ as $a \to \infty$.

Lemma. *If $\gamma(t) \in L$ and $\gamma(t)$ is continuous at 0 and $\gamma(t) \sim \int e^{ixt} g(x)\,dx$ and $g(x) \geqslant 0$ for all x, then $g \in L$ and $\int g(x) = \gamma(0)$.*

By the definition of $g(x)$,

$$\begin{aligned}
\int_{-b}^{b}\left(1-\frac{|x|}{b}\right) g(x)\,dx &= \frac{1}{2\pi}\int_{-b}^{b}\left(1-\frac{|x|}{b}\right) dx \int e^{-ixt}\gamma(t)\,dt \\
&= \frac{1}{2\pi}\int \gamma(t)\,dt \int_{-b}^{b}\left(1-\frac{|x|}{b}\right) e^{-ixt}\,dx \\
&= \frac{1}{\pi}\int \gamma\left(\frac{2t}{b}\right)\left(\frac{\sin t}{t}\right)^2 dt
\end{aligned}$$

and the conclusion follows on letting $b \to \infty$, since $\gamma(2t/b) \to \gamma(0)$ for all t,

$1-|x|/b \to 1$ for all x, and $g(x) \in L$ since $g(x) \geqslant 0$ and

$$\int_{-b/2}^{b/2} g(x)\,dx \leqslant 2\int_{-b}^{b}\left(1-\frac{|x|}{b}\right)g(x)\,dx,$$

which is bounded.

We complete the proof of the theorem by taking $\gamma(x) = \int \phi(x-u)\overline{\phi(-u)}\,du$ in the Lemma, for γ is integrable by Theorem 11(i) and continuous by Theorem 12. Moreover, $\gamma(t) \sim \int e^{ixt} g(x)\,dx$, where

$$\begin{aligned} g(x) &= \frac{1}{2\pi}\int e^{-ixt}\gamma(t)\,dt = \frac{1}{2\pi}\int e^{-ixt}\,dt \int \phi(t-u)\overline{\phi(-u)}\,du \\ &= \frac{1}{2\pi}\int \overline{\phi(-u)}\,du \int e^{-ixt}\phi(t-u)\,dt \\ &= \frac{1}{2\pi}\int e^{-ixu}\overline{\phi(-u)}\,du \int e^{-ixt}\phi(t)\,dt \\ &= 2\pi f(x)\overline{f(x)} = 2\pi\,|f(x)|^2 \end{aligned}$$

and it follows that $f \in L_2$ and $2\pi\,\|f\|_2^2 = \gamma(0) = \|\phi\|_2^2$.

Finally, we have

$$f(x) - f_a(x) = \frac{1}{2\pi}\left[\int_{-\infty}^{-a} + \int_{a}^{\infty}\right] e^{-ixt}\phi(t)\,dt,$$

and the first part of the theorem already proved, and applied to the function ϕ modified by reducing to zero its values in the interval $(-a, a)$, shows that

$$2\pi\,\|f-f_a\|_2^2 = \left[\int_{-\infty}^{-a} + \int_{a}^{\infty}\right] |\phi(t)|^2\,dt = o(1)$$

as $t \to \infty$.

This result can now be used to take a further step and drop the condition that $\phi \in L$.

Theorem 22. *If $\phi \in L_2$ and $f_a(y) = (1/2\pi)\int_{-a}^{a} e^{-ixt}\phi(t)\,dt$, there is a unique function f of L_2 such that $\|f_a - f\|_2 \to 0$ as $a \to \infty$, and $2\pi\,\|f\|_2^2 = \|\phi\|_2^2$.*

If $a' > a > 0$, $f_{a'} - f_a$ is the transform of the function with values $\phi(t)$ for $a \leqslant |t| \leqslant a'$ and zero elsewhere. This function is in $L \cap L_2$ and so, by Theorem 21,

$$2\pi\,\|f_{a'} - f_a\|_2^2 = \left[\int_{-a'}^{-a} + \int_{a}^{a'}\right] |\phi(t)|^2\,dt$$

and since this tends to 0 as $a, a' \to \infty$, a unique function f is defined by

Theorem 37 of Chapter 2 so that $\|f_a - f\|_2 \to 0$ as $a \to \infty$, and

$$2\pi \|f\|_2^2 = \lim_{a\to\infty} \|f_a\|_2^2 = \lim_{a\to\infty} \int_{-a}^{a} |\phi(t)|^2 \,dt = \|\phi\|_2^2.$$

We call f the **Fourier transform** of ϕ and write $\phi(t) \sim \int e^{ixt} f(x)\,dx$ and note that this terminology is consistent with that used for functions ϕ of L since the last theorem shows that the two definitions of f coincide when $\phi \in L \cap L_2$.

We can now establish the celebrated theorem of Plancherel which asserts the symmetrical relationship between a function of L_2 and its transform.

Theorem 23 (Plancherel). *If $\phi \in L_2$ and $\phi(t) \sim \int e^{ixt} f(x)\,dx$, then $f \in L_2$ and $f(x) \sim (2\pi)^{-1} \int e^{-ixt} \phi(t)\,dt$. (In other words $\bar{\phi}$ is the transform of $2\pi \bar{f}$.)*

It follows from Theorem 22, since $f \in L_2$, that f has a Fourier transform in L_2 and we can write

$$f(x) \sim \frac{1}{2\pi} \int e^{-ixt} \psi(t)\,dt, \qquad \psi_a(t) = \int_{-a}^{a} e^{ixt} f(x)\,dx,$$

$$\|\psi_a - \psi\|_2 \to 0 \quad \text{as} \quad a \to \infty.$$

We can also write

$$f_a(x) = \frac{1}{2\pi} \int_{-a}^{a} e^{-ixt} \phi(t)\,dt, \qquad \|f_a - f\|_2 \to 0 \quad \text{as} \quad a \to \infty$$

by the definition of f. We have to prove that $\psi(t) = \phi(t)$ a.e. We introduce two auxiliary functions, γ and its transform g, both in $L \cap L_2$, so that $\gamma(t) = \int e^{ixt} g(x)\,dx$, $g(x) = (2\pi)^{-1} \int e^{-ixt} \gamma(t)\,dt$. Then for every $a > 0$,

$$\begin{aligned} \int_{-a}^{a} \gamma(t-u)\phi(u)\,du &= \int_{-a}^{a} \phi(u)\,du \int e^{ix(t-u)} g(x)\,dx \\ &= \int e^{ixt} g(x)\,dx \int_{-a}^{a} e^{-ixu} \phi(u)\,du \\ &= 2\pi \int e^{ixt} g(x) f_a(x)\,dx, \end{aligned}$$

and since the integrand on the left is integrable in $(-\infty, \infty)$ as γ, ϕ both belong to L_2, and since $f_a \to f$ in L_2, we can let $a \to \infty$ and obtain

$$\int \gamma(t-u)\phi(u)\,du = 2\pi \int e^{ixt} g(x) f(x)\,dx. \tag{3}$$

On the other hand,

$$\int \gamma(t-u)\psi_a(u)\,du = \int \gamma(t-u)\,du \int_{-a}^{a} e^{ixt}f(x)\,dx$$
$$= \int_{-a}^{a} e^{ixt}f(x)\,dx \int e^{-ix(t-u)}\gamma(t-u)\,du$$
$$= 2\pi \int_{-a}^{a} e^{ixt}f(x)g(x)\,dx$$

and the integrand on the right is integrable, $\gamma \in L_2$ and $\psi_a \to \psi$ in L_2, and so

$$\int \gamma(x-u)\psi(u)\,du = 2\pi \int e^{ixt}f(x)g(x)\,dx.$$

Combining this with (3) and writing $\tau = \psi - \phi$, we have $\int \gamma(t-u)\tau(u)\,du = 0$. In particular, if we specify γ and its transform g by the 'triangular function' with

$$\gamma(t) = \max\{h^{-1}[1-|t|\,h^{-1}], 0\}, \qquad \pi g(x) = 2[(hx)^{-1}\sin(hx/2)]^2,$$

it is obvious that γ and g both belong to $L \cap L_2$ and we have, with $T(u) = \int_0^u \tau(v)\,dv$ and using partial integration,

$$\int_0^1 (hv)^{-1}[T(t+hv) - T(t-hv)]2v\,dv = h^{-2}\int_0^h [T(t+u) - T(t-u)]\,du$$
$$= \int T(t-u)\gamma'(u)\,du$$
$$= \int \tau(t-u)\gamma(u)\,du = 0.$$

The left-hand side tends to $T'(t) = \tau(t)$ for almost all t, by Theorem 18 of Chapter 2, and so $\tau(t) = 0$ a.e. as required.

We end this section with the analogues for transforms of the convolution Theorems 14 and 16 already proved for series, noting first that, by Theorem 11(i), $\phi * \gamma \in L$ if $\phi \in L$ and $\gamma \in L$.

Theorem 24. *If $\phi, \gamma \in L$ and*

$$\phi(t) \sim \int e^{ixt}f(x)\,dx; \qquad \gamma(t) \sim \int e^{ixt}g(x)\,dx,$$

then

$$\phi * \gamma(t) \sim 2\pi \int e^{ixt}g(x)f(x)\,dx,$$

for

$$\int e^{-ixt}\phi*\gamma(t)\,dt=\int e^{-ixt}\,dt\int\gamma(u)\phi(t-u)\,du$$
$$=\int\gamma(u)\,du\int e^{-ixt}\phi(t-u)\,dt$$
$$=\int e^{-ixu}\gamma(u)\,du\int e^{-ix(t-u)}\phi(t-u)\,dt$$
$$=\int e^{-ixu}\gamma(u)\,du\int e^{-ixt}\phi(t)\,dt=4\pi^2 g(x)f(x),$$

as required.

We note, of course, that the theorem says nothing about the convergence of the Fourier integral representation of $\phi*\gamma$, but the inherent lack of symmetry disappears in the L_2 case. We note that, after Theorem 12, $\phi*\gamma(t)$ exists for all t if $\phi,\gamma\in L_2$.

Theorem 25. *If $\phi, \gamma\in L_2$ and have transforms f, g, then*

$$\phi*\gamma(t)=2\pi\int e^{ixt}f(x)g(x)\,dx.$$

We define ϕ_a, g_b as in the previous theorems by

$$\phi_a(t)=\int_{-a}^{a} e^{ixt}f(x)\,dx, \qquad g_b(x)=\frac{1}{2\pi}\int_{-b}^{b} e^{-ixu}\gamma(u)\,du,$$

so that

$$2\pi\int_{-a}^{a} e^{ixt}f(x)g_b(x)\,dx=\int_{-a}^{a} e^{ixt}f(x)\,dx\int_{-b}^{b} e^{-ixu}\gamma(u)\,du$$
$$=\int_{-b}^{b}\gamma(u)\,du\int_{-a}^{a} e^{ix(t-u)}f(x)\,dx$$
$$=\int_{-b}^{b}\gamma(u)\phi_a(t-u)\,du$$

If we now let $a\to\infty$ and use the fact that $\phi_a\to\phi$ in L_2 and $\gamma\in L_2$, we get

$$2\pi\int e^{ixt}f(x)g_b(x)\,dy=\int_{-b}^{b}\gamma(u)\phi(t-u)\,du$$

and finally, letting $b\to\infty$ and noting that $f\in L_2$, $g_b\to g$ in L_2 and $\gamma(u)\phi(t-u)\in L$, the conclusion follows.

5.5 Fourier–Stieltjes transforms

The last case of harmonic representation we discuss is that in which F in equation (1) is not necessarily absolutely continuous or defining a measure concentrated at integer points as when ϕ is periodic. To compensate for this substantial formal generalization, we impose a very heavy restriction on F: that it should be of bounded variation in the whole interval $(-\infty, \infty)$. This can be expressed as $\int |dF| < \infty$ and it is then plain that the function ϕ defined by

$$\phi(t) = \int e^{ixt}\, dF(x), \qquad \|\phi\| = \int |dF| < \infty$$

exists for all values of t and is bounded. If F were also absolutely continuous, the case would revert, of course to that of the Fourier transform of a function of L (but with the roles of ϕ and f interchanged).

There is no problem of convergence, of course, since this is absolute, but there is a major problem in any attempt to characterize the class A of functions ϕ which can be expressed in this way. However, this class is of such fundamental importance in probability, economics and many branches of science that it is useful and necessary to know something about the properties of A as a class. We deal only with the most familiar of these. Since the measure defined by F is not affected by its values at points of discontinuity, there is no loss of generality in supposing that F is normalized so that $F(x) = [F(x+0) + F(x-0)]/2$.

Theorem 26 (Inversion formula). *If* $\phi(t) = \int e^{ixt}\, dF(x)$, $\|\phi\| = \int |dF| < \infty$, *then*

$$F(x) - F(a) = \lim_{A\to\infty} \int_{-A}^{A} \frac{e^{-ixt} - e^{-iat}}{-2\pi i t}\, \phi(t)\, dt,$$

provided that $F(y)$ *is normalized.*

By Theorem 33 of Chapter 2, we have

$$\begin{aligned}\int_{-A}^{A} (e^{-ixt} - e^{-iat}) t^{-1} \phi(t)\, dt &= \int_{-A}^{A} (e^{-ixt} - e^{-iat}) t^{-1}\, dt \int e^{iut}\, dF(u)\\ &= 2i \int dF(u) \int_0^{A} [\sin t(u-x) - \sin t(u-a)] t^{-1}\, dt\\ &= 2i \int \{S[A(u-x)] - S[A(u-a)]\}\, dF(u),\end{aligned}$$

$$S(\xi) = \int_0^{\xi} t^{-1} \sin t\, dt.$$

It is familiar that $S(\xi)$ is bounded and $S(\xi)\to\pm\pi/2$ according to $\xi\to\pm\infty$, and it follows that the $S[A(u-x)]-S[A(u-a)]$ is bounded and tends to $-\pi/2$ for $u=x$ or $u=a$, to $-\pi$ for $a<u<x$ and to 0 for u outside the interval $[a,x]$. The conclusion follows from Theorem 18 of Chapter 2.

Corollary. *Under the conditions of the theorem, F is uniquely determined by ϕ except possibly at its discontinuities.*

Theorem 27. *If $\phi\in A$, then ϕ is bounded and therefore uniformly continuous.*

This follows from

$$|\phi(t)|=\left|\int e^{ixt}\,dF(x)\right|\leqslant\int|dF(x)|=\|\phi\|,$$

and if $h>0$,

$$|\phi(x+h)-\phi(t)|=\left|\int e^{ixt}[e^{ixt}-1]\,dF(x)\right|\leqslant\int|e^{ixh}-1|\,dF$$

and the conclusion follows from Theorem 18 of Chapter 2.

Theorem 28. *If ϕ, γ belong to A and a, b are constants, then $a\phi+b\gamma\in A$ and $\|a\phi+b\gamma\|\leqslant|a|\,\|\phi\|+|b|\,\|\gamma\|$.*

In other words, A is a linear metric space with $\|\phi\|$ as metric. In fact, Theorem 33 below shows that $\phi\gamma$ also belongs to A, which is therefore an algebra. The proof of Theorem 28 is obvious.

It is sometimes interesting to extend the Radon–Nikodym decomposition of F by separating out its discontinuities c_n at points λ_n. We then have

$$\phi(t)=\int e^{ixt}f(x)\,dx+\sum c_n e^{i\lambda_n t}+\int e^{ixt}\,dS(x),$$

where $f(x)=F'(x)\in L$, $\sum|c_n|<\infty$, $\int|dS|<\infty$ and S is continuous, $S'(x)=0$ a.e. The function S is not necessarily constant and is the continuous singular component of F. In fact, it is sometimes called *the* singular component, but this conflicts slightly with our earlier use of the word singular to include the discontinuous component as well as S. The case in which $F'=0$, $Q=0$ has analogues with the case of absolutely convergent Fourier series, to which it reduces when all λ_n are integral multiples of a common base. Except in the Fourier case, ϕ is not periodic, but is almost periodic in a sense which can be expressed in terms of its differences $\phi(t+l)-\phi(t)$; and extensions of the notion of almost periodicity can be made which do not depend on the absolute convergence of $\sum c_n$. For functions of A, however, there is a simple theorem which is an extension of the definition of Fourier coefficients in the periodic case.

Theorem 29. *If $\phi \in A$ and F has a discontinuity c at λ, then*

$$c = \lim_{T\to\infty} \frac{1}{2T} \int_{-T}^{T} e^{-i\lambda t}\phi(t)\, dt.$$

This is shown by

$$\begin{aligned}\frac{1}{2T}\int_{-T}^{T} e^{-i\lambda t}\phi(t)\, dt &= \frac{1}{2T}\int_{-T}^{T} e^{-i\lambda t}\, dt \int e^{ixt}\, dF(x)\\ &= \int dF(x) \frac{1}{2T}\int_{-T}^{T} e^{it(x-\lambda)}\, dt\\ &= \int \frac{\sin T(x-\lambda)}{T(x-\lambda)}\, dF(x)\\ &= c + \int \frac{\sin T(x-\lambda)}{T(x-\lambda)}\, dF^*(x),\end{aligned}$$

where $\int |dF^*| < \infty$ and F^* is continuous at λ. Since $\sin T(x-\lambda)[T(x-\lambda)]^{-1}$ is bounded in $(-\infty, \infty)$ and tends to 0 as $T \to \infty$ at every point except λ, and so almost everywhere with respect to the measure F^*, the conclusion follows.

In the remainder of this section we deal with properties of convolutions of function F of bounded variation and note that there is no loss of generality in supposing that they are non-decreasing and that $F(-\infty) = 0$, $F(+\infty) = 1$, so that $\int dF = 1$. Such a function is called a (probability) **distribution function** and the associated function ϕ is called its **characteristic function,** and both are of fundamental importance in the theory of probability.

Theorem 30. *If F_1, F_2 are distribution functions, so is $F = F_1 * dF_2$ defined by*

$$F_1 * dF_2(x) = \int F_1(x-y)\, dF_2(y) = \int F_2(x-y)\, dF_1(y).$$

First, it is obvious that $F(x)$ increases and, if a is any point of continuity of F_2, we have

$$F(x) \geq \int_{-\infty}^{a} F_1(x-y)\, dF_2(y) \geq F_1(x-a)F_2(a)$$

$$\begin{aligned}F(x) &= \int_{-\infty}^{a} F_1(x-y)\, dF_2(y) + \int_{a}^{\infty} F_1(x-y)\, dF_2(y)\\ &\leq F_2(a) + F_1(x-a)[1 - F_2(a)].\end{aligned}$$

Since $F_1(-\infty) = F_2(-\infty) = 0$ and $F_1(\infty) = F_2(\infty) = 1$, this gives $0 \leq F(-\infty) \leq F_2(a) \leq F(\infty) \leq 1$ for every a, and so $F(-\infty) = 0$, $F(\infty) = 1$.

Theorem 31. *If F_1, F_2 are distribution functions and F_1 is absolutely continuous with derivative f_1, then $F = F_1 * \mathrm{d}F_2$ is absolutely continuous and $f(x) = F'(x) = \int f_1(x-y)\,\mathrm{d}F_2(y)$ a.e.*

If F_2 is also absolutely continuous with derivative f_2, then

$$f(x) = \int f_1(x-y) f_2(y)\,\mathrm{d}y.$$

We have

$$F(x) = \int F_1(x-y)\,\mathrm{d}F_2(y) = \int \mathrm{d}F_2(y) \int_{-\infty}^{x-y} f_1(u)\,\mathrm{d}u$$

by Theorem 11 of Chapter 3.

$$= \int \mathrm{d}F_2(y) \int_{-\infty}^{x} f_1(u-y)\,\mathrm{d}u = \int_{-\infty}^{x} \mathrm{d}u \int f_1(u-y)\,\mathrm{d}F_2(y)$$

by Fubini's Theorem (Theorem 33 of Chapter 2) since $f_1(u-y)$ is Borel measurable with respect to the product of F_2 and ordinary Lebesgue measure. The conclusion follows from Theorem 11 of Chapter 3 and the second part follows from Theorem 30 of Chapter 2.

Theorem 32. *Suppose that F_1, F_2 are distribution functions, $F = F_1 * \mathrm{d}F_2$ and α is a bounded Borel function. Then $\int \alpha(x+y)\,\mathrm{d}F_2(y)$ is integrable with respect to $F_1(y)$ and $\int \mathrm{d}F_1(y) \int \alpha(x+y)\,\mathrm{d}F_2(x) = \int \alpha(x)\,\mathrm{d}F(x)$.*

Suppose first that $\alpha(x)$ is the characteristic function of the interval $a \leqslant x < b$. The left-hand side is then

$$\int \mathrm{d}F_1(y) \int_{a-y-0}^{b-y-0} \mathrm{d}F_2(x) = \int [F_2(b-y-0) - F_2(a-y-0)]\,\mathrm{d}F_1(y)$$

$$= F(b-0) - F(a-0) = \int \alpha\,\mathrm{d}F.$$

The theorem therefore holds in this case and extends immediately to simple functions and, after Theorem 20 of Chapter 2, to countable sums of non-negative simple functions. We can define a sequence of simple functions θ_n so that $|\alpha - \theta_n| \leqslant \lambda_n$, λ_n is a countable sum of non-negative simple functions and $\int \lambda_n\,\mathrm{d}F \to 0$, $\int \theta_n\,\mathrm{d}F \to \int \alpha\,\mathrm{d}F$.

It follows from this and Fubini's theorem, since λ_n is measurable with respect to the product measure of F_1 and F_2, that

$$\iint \lambda_n(x+y)\,\mathrm{d}F_1\,\mathrm{d}F_2 = \int \lambda_n\,\mathrm{d}F \to 0.$$

But $\iint |\alpha(x+y) - \theta_n(x+y)|\,\mathrm{d}F_1\,\mathrm{d}F_2 \leqslant \iint \lambda_n(x+y)\,\mathrm{d}F_1\,\mathrm{d}F_2 = o(1)$, and $\alpha(x+y)$ is integrable and measurable with respect to the product measure by Theorems 9 and 22 of Chapter 2. We can therefore use Fubini's Theorem again, and deduce that

$$\int dF_1(y)\int \alpha(x+y)\,dF_2(x)=\iint \alpha(x+y)\,dF_1\,dF_2$$

$$=\lim\iint \theta_n(x+y)\,dF_1\,dF_2$$

$$=\lim\int \theta_n\,dF=\int \alpha\,dF.$$

Theorem 33. *If ϕ_1, ϕ_2 are the characteristic functions of distribution functions F_1 and F_2, respectively, then $\phi_1\phi_2$ is the characteristic function of $F=F_1*dF_2$.*

This follows from

$$\phi_1(t)\phi_2(t)=\int e^{iyt}\,dF_1(y)\int e^{ixt}\,dF_2(x)$$

$$=\int dF_1(y)\int e^{it(x+y)}\,dF_2(x)=\int e^{ixt}\,dF(x)$$

by Theorem 32, with $\alpha(x)=e^{ixt}$.

5.6 Spectra and filters

The idea of a spectrum of a function ϕ is inherent in the basic formula $\phi(t)\sim\int e^{itx}\,dF(x)$, for this expresses a signal ϕ as an integral or sum of pure wave forms e^{itx} so that F can be thought of as defining the spectrum of ϕ. The inversion formula indicates the basic duality between ϕ and F which takes its most symmetrical form in the case of Fourier transforms when ϕ and $f=F'$ both belong to L_2. The harmonic representation of ϕ in this way has many physical interpretations when $\phi(t)$ is the value of some familiar quantity at time t. If ϕ measures the electric current through a circuit of constant resistance, for example, $|\phi|^2$ defines the power being transmitted and its integral over a time interval gives the total energy dispersed over this time. The most important special cases are those in which ϕ is either periodic and $L_2(-\pi,\pi)$ or $\phi\in L_2(-\infty,\infty)$. In the first case, the energy of the signal ϕ over each interval of length 2π is, by Parseval's Theorem, the sum of the energies $|c_n|^2$ from the separate components c_ne^{int} of its Fourier series. In the second case, the total energy of the signal over the whole time interval $(-\infty,\infty)$ is 2π times the integral of the power $|f(x)|^2$ of the simple wave $e^{itx}f(x)$.

This familiar idea of a spectrum deriving from classical physics has, of course, been generalized widely over the last few decades and it is not possible here to comment usefully on its application to unitary operators in Hilbert spaces and elsewhere except to say that these are consistent conceptually with the earlier and more immediately applicable development. The relationship

$$\phi(t\lambda) = \lambda^{-1} \int e^{itx} f(x\lambda^{-1})\, dx$$

suggests that a signal and its spectrum cannot both be concentrated near the origin. It is not easy to formulate this principle precisely, but it is of great practical importance in electrical engineering and communication theory and there are many results dealing with particular cases of it. The most famous is the following theorem which forms the mathematical basis of Heisenberg's uncertainty principle.

Theorem 34. *Suppose that $\phi \in L_2$, $\phi(t) \sim \int e^{itx} f(x)\, dx$ (so that $f \in L_2$) and that xf, $t\phi$ also belong to L_2. Then*

$$\|t\phi\|_2 \, \|xf\|_2 \geqslant (8\pi)^{-1/2} \, \|\phi\|_2^2.$$

(If we suppose that $\|\phi\|_2 = 1$, $|\phi|^2$ and $2\pi\,|f|^2$ are probability density functions with standard deviations σ, s, and the conclusion can be written $\sigma s \geqslant (8\pi)^{-1/2}$. The theorem applies in this form to quantum physics with $|\phi|^2$, $2\pi\,|f|^2$ the probability densities of the position and momentum of a particle in 'state' ϕ.)

We prove the theorem first in the case in which $f(x) = 0$ outside a finite interval and f has a bounded second derivative. This is enough to ensure that, by Theorem 9 of Chapter 3,

$$\phi'(t) = i \int e^{itx} x f(x)\, dx, \qquad \|\phi'\|_2^2 = 2\pi\, \|xf\|_2^2.$$

Using Theorem 34 (Schwartz) of Chapter 2,

$$\begin{aligned}
(8\pi)^{1/2} \|t\phi\|_2 \, \|xf\|_2 = 2\, \|t\phi\|_2 \, \|\phi'\|_2 &\geqslant 2 \int |t\phi(t)\phi'(t)|\, dt \\
&\geqslant \int |t[\phi(t)\bar{\phi}'(t) + \bar{\phi}(t)\phi'(t)]|\, dt \\
&= \int |t(|\phi|^2)'|\, dt \\
&\geqslant \left| \int t(|\phi|^2)'\, dt \right| = \|\phi\|_2^2
\end{aligned}$$

by partial integration. The proof is completed if we observe that f, xf, ϕ, $t\phi$ can always be approximated simultaneously in L_2 by functions of the types just considered.

The definition of a convolution enables us to introduce the idea of a **filter**, and this is illustrated most simply in the case of a periodic function ϕ with Fourier series $\sum c_n e^{int}$. If γ is a trigonometric polynomial $\sum d_n e^{int}$ with only a finite number of terms, we have

$$\psi(t) = \int \phi(t-u)\gamma(u)\,du \sim 2\pi \sum c_n d_n e^{int}$$

and ψ then represents the result of filtering out from ϕ all the frequencies for which $d_n = 0$. In the Fourier transform case

$$\psi(t) = \int \phi(t-n)\gamma(u)\,du \sim 2\pi \int e^{iyx} f(x)g(x)\,dx$$

and ψ is the signal remaining after the frequencies for which $g(x)=0$ have been removed from ϕ. In practice, the requirement $g(x)=0$ for values of x is unnecessarily stringent and it is sufficient that $g(x)$ be small for certain values. Indeed there are many theorems about the limitations on what filters can do to modify a signal, and these have quite obvious physical meanings. This can be seen most obviously if the filtered signal γ is regarded as the result of adding values of ϕ delayed by backward time (or phase) shift u and magnified by a factor $\gamma(u)$. There are very well known ways of designing circuits to do this when $\phi(t)$ is an electric current, but the obvious fact that no immediate action can be taken on a future event means that $\gamma(u)=0$ for $u<0$. This physical restraint imposes, after a theorem of Hardy, the condition

$$\int (1+x^2)^{-1} \log |g(x)|\,dx > -\infty \quad \text{on } g,$$

so that $g(x)$ cannot be 'too small', for then $\log |g(x)|$ would be too large and negative; nor, in particular, can $g(x)=0$ throughout any interval.

Thinking of γ as itself a signal, this shows that a signal terminating at a certain time cannot be represented by a spectrum in which the frequency band is limited in either direction.

6
Random variables and probability

6.1 Definitions and basic ideas

The application of the foregoing theory of measure to the study of probability is based on the principle, first enunciated in its present form by Kolmogoroff,† that all statements involving the notion of probability can be expressed as statements about the measures of sets in an appropriate space.

The basic terms are **random variable** and **probability,** and we must either define them or give a precise meaning to every sentence in which they occur. All that is necessary if we adopt the second alternative is that we understand the two sentences

(i) x is a random variable in the space $\mathscr{X}$,
(ii) the probability that x belongs to X is p or, in symbols, $\mathbf{P}\{x \in X\} = p$

to mean that a **probability measure** (or **probability distribution**) μ is defined in $\mathscr{X}$, with the properties that $\mu(X) \geqslant 0$, $\mu(\mathscr{X}) = 1$, that X is measurable and that $\mu(X) = p$. There is no need to deal with the sentences (i) and (ii) separately since it is invariably found that (i) is followed by one or more statements, which may be hypotheses or conclusions, of type (ii). In fact, from this point of view it is not necessary to go any further and say what a random variable *is*, any more than it is necessary to define the isolated symbol ∞ in analysis.

This has been done, however, by Kolmogoroff in the case of random variables in the space $\mathscr{R}$ of real numbers. He says that a random real variable is a real valued function over a space $\mathscr{X}$ which is measurable with respect to a probability measure in $\mathscr{H}$. After Theorem 31 of Chapter 2, the measure in $\mathscr{X}$ induces a probability measure in $\mathscr{R}$ and this could equally well be used as the basic measure of the random variable, which would then become (in Kolmogoroff's usage) the identity function over $\mathscr{R}$. Conversely, a probability measure in $\mathscr{R}$ immediately defines through the identity function a random variable in Kolmogoroff's sense.

If we recall that the terms probability and random variable never occur in isolation, but only in the context described above, there is no need to distinguish between the two ways of introducing them. The one chosen here is conceptually simpler in that it does not need the idea of a measurable function or even a function.

† Kolmogoroff, A. (1956) *Foundations of the theory of probability*. Chelsea, New York.

When $\mathscr{X}$ is a particular space of a familiar kind, it is usually convenient to show this by indicating the nature of the random variable. Thus, we speak of a **random integer** when $\mathscr{X}$ is a set of integers, a **random real number** when $\mathscr{X}$ is $\mathscr{R}$ and a **random *k*-vector** when $\mathscr{H}$ is $\mathscr{R}^k$.

The probability measure of a random real number is expressed most simply in terms of its **distribution function** F, which is a real valued function on $\mathscr{R}$ whose values $F(x)$ increase and satisfy $F(-\infty)=0$, $F(\infty)=1$. We then have $\mathbf{P}\{x \leqslant c\} = F(c+0)$ and, more generally,

$$\mathbf{P}\{x \in X\} = \int_X \mathrm{d}F = F(X)$$

for any Borel set X. The distribution can also be defined by its **characteristic function** ϕ, defined by

$$\phi(t) = \int \mathrm{e}^{itx} \, \mathrm{d}F,$$

since this determines F uniquely by the corollary of Theorem 26 of Chapter 5.

We are concerned in this section with general concepts relating to random variables which do not depend on specific properties of the space $\mathscr{X}$, although we shall use special cases quite freely for illustration.

This formulation of the basic ideas of probability in terms of measure takes away the obscurity and mystification which have been associated with them in the past, for many ambiguities and paradoxes have arisen simply through the lack of clarity in the specification of the populations (probability spaces) under consideration. On the other hand, the establishment of a sound axiomatic basis for the mathematics of probability does not eliminate the genuine philosophical problems of the meaning of probability statements and arguments as applied to real situations. At the most rudimentary level, for example, it is plainly impossible to give precision to the sort of statement that 'in a large number' of tosses of a coin we 'expect about as many' heads as tails. Nevertheless all statisticians in practice have to make such statements or act in accordance with them, and it is not easy to explain objectively what is taking place. Indeed it is difficult to avoid a largely subjective element in the interpretation of assessments and decisions based on probability considerations. This view does not conflict with the measure theory approach provided that assumptions about measures made by any individual are consistent. In some situations such as gambling, for example, differences between odds (probability measures) accepted by different individuals are evident and characteristic. The more common situation, however, is that in which individuals are willing to agree on common assumptions, based on common experience, about the probability measures they are using, for this is the essential requirement for effective communication and cooperation. The acceptance of even odds in tossing a coin is still a subjective act notwithstanding the fact that an individual choosing a different probability and acting on it would be regarded as eccentric even by fellow statisticians.

Another advantage of admitting a subjective element is that it allows an individual decision to be made as to whether probability arguments are relevant or valid or not in a particular case. There was, for example, a long and quite sterile controversy about the validity of Bayes' Theorem and the concept of prior probabilities. This can be avoided through the common sense view that these concepts may be useful and have recognizable physical meaning in certain situations, but not valid or helpful in others; and that there is no logical inconsistency in leaving the decision on this to individuals.

Many problems in probability involve more than one random variable, and when these can be treated separately and without reference to one another, no new idea is required. But very often we have to consider them together within a common framework, and rather more care is then needed. We say that $x_1, x_2, \ldots, x_k$ are random variables over spaces $\mathscr{X}_1, \mathscr{X}_2, \ldots, \mathscr{X}_k$, respectively, if $x = (x_1, x_2, \ldots, x_k)$ is a random variable in their product space $\mathscr{X} = \mathscr{X}_1 \times \mathscr{X}_2 \ldots \mathscr{X}_k$. The probability distribution $\mu(X)$ of x in $\mathscr{X}$ is called the **joint probability distribution** of $x_1, x_2, \ldots, x_k$. The sets X_1 of $\mathscr{X}_1$ for which $X_1 \times \mathscr{X}_2 \ldots \mathscr{X}_k$ is measurable with respect to μ clearly form a σ-ring in $\mathscr{X}_1$ on which there is a probability $\mu_1(X_1)$ defined by

$$\mu_1(X_1) = \mu(X_1 \times X_2 \ldots X_k),$$

and we can therefore say that x_1 is a random variable in $\mathscr{X}_1$ with probability distribution μ_1. In the same way, $x_2, \ldots, x_k$ are random variables in $\mathscr{X}_2, \ldots, \mathscr{X}_k$, with distributions $\mu_2, \ldots, \mu_k$, respectively.

If μ is the product measure of measures in $\mathscr{X}_1, \ldots, \mathscr{X}_k$, these measures must clearly be $\mu_1, \mu_2, \ldots, \mu_k$, and we say then that the random variables $x_1, \ldots, x_k$ are **independent**. Otherwise, they are called **dependent**. The number of variables is assumed here to be finite. The extension to infinite sets of variables involves deeper ideas and is postponed until later.

There is no difficulty in giving examples of independent or dependent variables. The probability distribution over the unit square $0 \leqslant x_1 \leqslant 1$, $0 \leqslant x_2 \leqslant 1$, in which $\mu(X)$ for a Borel set X is its ordinary Lebesgue measure, clearly satisfies the condition for independence. On the other hand, the random variables (x_1, x_2) which can take only values 0 and 1 are not independent if their joint probability distribution is defined by

$$\mathbf{P}\{x_1 = 0, x_2 = 0\} = 0, \qquad \mathbf{P}\{x_1 = 1, x_2 = 1\} = \tfrac{1}{2},$$
$$\mathbf{P}\{x_1 = 0, x_2 = 1\} = \mathbf{P}\{x_1 = 1, x_2 = 0\} = \tfrac{1}{4}.$$

We have seen in Section 2.3 that if a function α with values $y=\alpha(x)$ maps $\mathscr{X}$ onto $\mathscr{Y}$, the sets Y of $\mathscr{Y}$ which are inverse images of sets X which are measurable with respect to a probability measure μ in $\mathscr{X}$ form a σ-ring in $\mathscr{Y}$, and we get a probability distribution ν in $\mathscr{Y}$ by defining $\nu(Y)=\mu(X)$ for the (ν-measurable) sets Y for which $Y=\alpha^{-1}(X)$ and X is μ-measurable. We therefore speak of a random variable y as the value $\alpha(x)$ of the **function** α **of a random variable** x to mean that

$$\mathbf{P}\{y \quad Y\}=\nu(Y)=\mu(X).$$

In order that the σ-ring of measurable sets in y is not trivial, it is usually necessary to impose some conditions on the function α. In the particularly important case of a real valued function for which $\mathscr{Y}=\mathscr{R}$, the measurability of $\alpha(x)$ in the sense of Section 2.3 is enough to ensure that $\nu(Y)$ is defined at least for all Borel sets Y.

In the following examples we suppose that $\mathscr{X}=\mathscr{Y}=\mathscr{R}$ and that x has distribution function F and characteristic function ϕ, and we determine the distribution and characteristic functions G, ψ of $y=\alpha(x)$.

Example 1. $y=\alpha(x)=x^2$.

Since $y\geqslant 0$ for all x, it is plain that $G(y)=0$ for $y<0$. If $y\geqslant 0$, we must have

$$\begin{aligned}G(u+0)=\mathbf{P}\{y\leqslant u\}=\mathbf{P}\{x^2\leqslant u\}&=\mathbf{P}\{-u^{1/2}\leqslant x\leqslant u^{1/2}\}\\&=F(u^{1/2}+0)-F(-u^{1/2}-0),\end{aligned}$$

and since the probability measure defined by G is independent of its values at discontinuities, we can define

$$G(y)=F(y^{1/2})-F(-y^{1/2}),$$

and it then follows that

$$\psi(t)=\int e^{ity}\,dG(y)=\int_0^\infty e^{ity}\,d[F(y^{1/2})-F(-y^{1/2})]=\int e^{itx^2}\,dF(x).$$

Example 2. $y=\alpha(x)=x^{-1}$.

Since y must be defined with probability 1, we must have $\mathbf{p}\{x=0\}=0$ and F must be continuous at 0. Moreover, $y\neq 0$ for real x and so $\mathbf{P}\{y=0\}=0$. If $u<0$,

$$G(u+0)=\mathbf{P}\{y\leqslant u\}=\mathbf{P}\{x^{-1}\leqslant u\}=\mathbf{P}\{u^{-1}\leqslant x<0\}=F(0)-F(u^{-1}-0),$$

and if $u>0$,

$$\begin{aligned}G(u+0)&=\mathbf{P}\{y<0\}+\mathbf{P}\{0<y\leqslant u\}=\mathbf{P}\{x<0\}+\mathbf{P}\{x\geqslant u^{-1}\}\\&=F(0)-1-F(u^{-1}-0),\end{aligned}$$

and we can therefore write

$$G(y)=F(0)-F(y^{-1}) \qquad \text{or} \qquad G(y)=F(0)+1-F(y^{-1})$$

according as $y<0$ or $y>0$. Also,

$$\psi(t)=-\int_{-\infty}^{0} e^{ity}\, dF(y^{-1})-\int_{0}^{\infty} e^{ity}\, dF(y^{-1})$$

$$=\int_{0}^{\infty} e^{itx^{-1}}\, dF(x)+\int_{-\infty}^{0} e^{itx^{-1}}\, dF(x)=\int e^{itx^{-1}}\, dF(x).$$

Example 3. $y=\alpha(x)=Ax+B$.

$$G(y+0)=\mathbf{P}\{Ax+B\leqslant y\}=\mathbf{P}\{x\leqslant (y-B)/A\}=F[(y-B)/A+0]$$

if $A>0$,

$$G(y+0)=\mathbf{P}\{Ax+B\leqslant y\}=\mathbf{P}\{x\geqslant (y-B)/A\}=1-F[(y-B)/A-0]$$

if $A<0$; and we can write

$$G(y)=F[(y-B)/A] \qquad \text{or} \qquad 1-F[y-B)/A]$$

according as $A>0$ or $A<0$. In both cases it follows easily that $\psi(t)=e^{itB}\phi(At)$.

The idea of several random variables $y_1, \ldots, y_k$ defined simultaneously as values $y_1=\alpha_1(x), \ldots, y_k=\alpha_k(x)$ of functions of the same random variable x contains nothing essentially new. We merely write $y=(y_1, y_2, \ldots, y_k)=[\alpha_1(x), \ldots, \alpha_k(x)]$ to define the mapping α of X onto the product space $\mathscr{Y}=\mathscr{Y}_1\times\ldots\times\mathscr{Y}_k$. The distribution of $y_1, y_2, \ldots, y_k$, and questions of dependence or independence can be treated in the way described above.

We can now introduce two important parameters associated with a random real variable. The first, called the **mean** or **expectation** of random variable x with distribution function F is denoted by $\mathbf{E}\{x\}$ and defined by

$$\mathbf{E}\{x\}=\int x\, dF$$

whenever the integral exists in the Lebesgue sense The definition extends readily, as the following theorem shows, to the mean of a real or complex valued function of a random variable in any space.

Theorem 1. *If α is a measurable real or complex valued function of a random variable x with distribution μ in a space $\mathscr{X}$, then*

$$\mathbf{E}\{y\}=\mathbf{E}\{\alpha(x)\}=\int_{\mathscr{X}} \alpha(x)\, d\mu$$

provided that the last integral exists.

The theorem simply says that

$$\int y\,\mathrm{d}G=\int_{\mathscr{X}}\alpha(x)\,\mathrm{d}\mu,$$

where G is the distribution function of y, and this follows immediately from Theorem 31 of Chapter 2. Under these circumstances, we say that α has a finite mean and use the notation $\mathbf{E}\{y\}$ only when this is the case.

If we replace $\alpha(x)$ by $e^{it\alpha(x)}$, we deduce immediately the following general result which has already been illustrated in the examples above.

Theorem 2. *The characteristic function ψ of a real-valued function α of a random variable x with distribution μ in a space $\mathscr{X}$ is given by*

$$\psi(t)=\mathbf{E}\{e^{it\alpha(x)}\}=\int e^{it\alpha(x)}\,\mathrm{d}\mu.$$

In particular, if x is a real variable with distribution function F, then

$$\psi(t)=\int e^{it\alpha(x)}\,\mathrm{d}F \textit{ and (the case } \alpha(x)=x)\ \phi(t)=\int e^{itx}\,\mathrm{d}F=\mathbf{E}\{e^{itx}\}.$$

The second parameter σ is called the **standard deviation** of x and its square called the **variance** of x. They are defined by

$$\sigma^2=\mathbf{E}\{(x-m)^2\},$$

where $m=\mathbf{E}\{x\}$.

Theorem 3. *If x is a random real variable*

$$\sigma^2=\mathbf{E}\{(x-m)^2\}=\int (x-m)^2\,\mathrm{d}F=\mathbf{E}\{x^2\}-m^2.$$

After Theorem 2, we need only observe that

$$\int (x-m)^2\,\mathrm{d}F=\int (x^2-2mx-x^2)\,\mathrm{d}F=\int x^2-2m\int x\,\mathrm{d}F+m^2\int \mathrm{d}F.$$

Theorem 4 (Tchebycheff's inequality). *If α is a non-negative real valued function of a random variable x in any probability space $\mathscr{X}$, and if $k>0$, then*

$$k\mathbf{P}\{\alpha(x)\geqslant k\}\leqslant \mathbf{E}\{\alpha(x)\}.$$

If μ is the probability distribution of x, we have

$$\mathbf{E}\{\alpha(x)\}=\int_{\mathscr{X}}\alpha(x)\,\mathrm{d}\mu\geqslant\int_{\alpha(x)\geqslant k}\alpha(x)\,\mathrm{d}\mu$$

$$\geqslant k\int_{\alpha(x)\geqslant k}\mathrm{d}\mu=k\mathbf{P}\{\alpha(x)\geqslant k\}.$$

The special case in which x is a random real variable and

$$\alpha(x) = (x-m)^2, \qquad k = \lambda^2\sigma^2$$

is worth stating separately.

Theorem 5. *If a random real variable x has mean m and standard deviation σ, and if $\lambda > 0$, then*

$$\mathbf{P}\{|x-m| \geqslant \lambda\sigma\} \leqslant \lambda^{-2}.$$

6.2 Random real numbers

The practical applications of probability to statistical and other problems are based predominantly on the theory of distributions in $\mathcal{R}^k$, and we shall devote this chapter to an account of this in the simplest and most familiar space $\mathcal{R}$ and deal rather more systematically with the properties of random real numbers and their distribution and characteristic functions.

Two special types are of particular interest. First, if a distribution function F is absolutely continuous, its derivative $f(x) = F'(x)$ exists almost everywhere and is called the **probability density** of the random variable.

In the second type, we have

$$F(x+0) = \sum_{\lambda_v \leqslant x} p_v, \qquad p_v > 0, \qquad \sum_{v=0}^{\infty} p_v = 1.$$

We say in this case that F is a **point distribution function.** It is plain that F is a step function if the points λ_v are isolated, and this is necessarily the case when their number is finite. But they need not be isolated, and in this case F is not a step function. It is possible, for example, for λ_v to include the dense set of rationals or any other countable set.

Theorem 6. *If x is a random real number with point distribution function F, then*

$$\mathbf{P}\{x = \lambda_v\} = p_v, \mathbf{P}\{x \neq \lambda_v \text{ for all } v\} = 0$$

Moreover, if α is a measurable function on $\mathcal{R}$, its values $\alpha(x)$ have finite mean

$$\mathbf{E}\{\alpha(x)\} = \sum_{v=0}^{\infty} p_v \alpha(\lambda_v)$$

provided that the series converges absolutely. In particular, the characteristic function ϕ of x is given by

$$\phi(t) = \sum_{v=0}^{\infty} p_v e^{it\lambda_v}.$$

We have

$$\mathbf{P}\{x=\lambda_v\}=F(\lambda_v+0)-F(\lambda_v-0)$$
$$=\lim_{\delta\to 0}\sum_{(\lambda_v-\delta<\lambda_j<\lambda_v+\delta)} p_j=p_v,$$

since every point λ_j other than λ_v is excluded from the interval $\lambda_v-\delta<x<\lambda_v+\delta$ when δ is sufficiently small. Since $\sum p_v=1$, it follows that $P\{x\neq\lambda_v$ for all $v\}=1-\sum p_v=0$, and after Theorem 1, it follows that

$$\mathbf{E}\{\alpha(x)\}=\int\alpha(x)\,\mathrm{d}F=\sum_{v=0}^{\infty}\int_{x=\lambda_v}\alpha(x)\,\mathrm{d}F=\sum_{v=0}p_v\alpha(\lambda_v).$$

In the practically important case in which x is a random integer and the numbers λ_v are all non-negative integers, we may suppose that $\lambda_v=v$ if we admit zero values of p_v when necessary. In this case,

$$\phi(t)=\sum_{v=0}^{\infty}p_v\mathrm{e}^{\mathrm{i}tv}=p(\mathrm{e}^{\mathrm{i}t}),$$

where p, defined for all complex z in the unit circle $|z|\leqslant 1$ by

$$p(z)=\sum_{v=0}^{\infty}p_v z^v,$$

is called the **probability generating function** of the distribution.

If we are dealing only with random integers, probability generating functions can be used instead of characteristic functions, and the theory can be based on the elementary properties of power series rather than the more elaborate analysis of Chapter 5.

If m is the mean of a random real number, we define μ_j, called the jth moment of x about its mean, or the jth central moment, by

$$\mu_j=\mathbf{E}\{(x-m)^j\},$$

so that $\mu_0=1$, $\mu_1=0$, $\mu_2=\sigma^2$, and the mean and moments can all be expressed immediately in terms of the distribution function.

It is also sometimes useful to express them as we do in the next two theorems in terms of the characteristic function or, in the case of random integers, the probability generating function.

Theorem 7. *If a random real number x has a finite mean, then its characteristic function ϕ is differentiable at $t=0$ and*

$$m=\mathbf{E}\{x\}=-\mathrm{i}\phi'(0).$$

If it has a finite variance, then ϕ has a second derivative at $t=0$ and

$$\sigma^2=\mu_2=[\phi'(0)]^2-\phi''(0).$$

If $h \neq 0$, we have

$$\phi(t+h)-\phi(t)=\int e^{itx}(e^{ihx}-1)\,dF,$$

and since $(e^{ihx}-1)/h \to ix$ as $h \to 0$ and $|e^{ihx}-1| \leqslant |hx|$ for all real x and h, the conclusion follows from Theorem 18 of Chapter 2. The second conclusion can be proved by a similar argument.

Theorem 8. *If n is a random integer with probability generating function p and finite mean m, then p has left derivative m at 1. If n has a finite variance, p has a left second derivative σ^2+m^2-m at 1. If p is once or twice differentiable at 1, the derivatives can be taken in the ordinary two-sided sense.*

If $m=\sum vp_v<\infty$ and $0 \leqslant \xi < 1$, we have

$$\lim_{\xi\to 1-0}\frac{p(1)-p(\xi)}{1-\xi}=\lim_{\xi\to 1-0}\sum_{v=0}^{\infty}p_v\frac{1-\xi^v}{1-\xi}=\sum_{v=0}^{\infty}vp_v=m,$$

since $(1-\xi^v)/(1-\xi)=1+\xi+\xi^2+\ldots+\xi^{v-1}$ increases and tends to v as $\xi \to 1-0$ and $\sum vp_v$ converges.

If the second moment is finite, $\sum v^2p_v<\infty$ and if $0 \leqslant \xi < 1$,

$$\lim_{\xi\to 1-0}\frac{p'(1)-p'(\xi)}{1-\xi}=\lim_{\xi\to 1-0}\sum_{v=0}^{\infty}vp_v\frac{(1-\xi^{v-1})}{1-\xi}$$

$$=\sum_{v=0}^{\infty}v(v-1)p_v=\sum_{v=0}^{\infty}v^2p_v-\sum_{v=0}^{\infty}vp_v=\sigma^2+m^2-m.$$

There are several other parameters of a random real number which are also used, though less frequently than the mean and standard deviation; and we touch on them only very briefly. First, a **median** of the random variable with distribution function F is any number q for which

$$F(q-0) \leqslant \tfrac{1}{2} \leqslant F(q+0).$$

A median always exists since the condition is obviously satisfied by the upper bound of the real members x for which $F(x) \leqslant \frac{1}{2}$, but it is not unique when $F(x)=\frac{1}{2}$ in an open interval $q_1<x<q_2$, for the median values there are the points q of the closed interval $q_1 \leqslant q \leqslant q_2$.

The number $\frac{1}{2}$ used in defining the median may be replaced by any number p in the range $0<p<1$, and a number q_p so defined that

$$F(q_p-0) \leqslant p \leqslant F(q_p+0)$$

is called the quantile of order p of the distribution. The values $p=\frac{1}{4}$, $p=\frac{3}{4}$ give upper and lower **quartiles,** and the interval $q_{1/4} \leqslant x \leqslant q_{3/4}$ is called the interquartile range. The quartiles, like the median, are not necessarily unique, but they often given a good account of the magnitude and spread of a random variable.

A distribution is said to have a finite **range** $[r, R]$ if $F(r-0)=0$, $F(R+0)=1$ and $0<F(x)<1$ for $r<x<R$, so that $P\{r\leqslant x\leqslant R\}=1$. In particular, a variable taking only a finite number of values has a range in which r is the least and R the greatest.

The **absolute moment** $\mathbf{E}\{|x-c|\}$ about a point c is occasionally used instead of the standard deviation as a measure of spread. The following theorem shows that it is related to the median, as is the standard deviation to the mean, by an interesting minimal property.

Theorem 9. (i) *The mean square deviation* $\mathbf{E}\{(x-c)^2\}$ *about a number* c *has the strict minimum value* σ^2 *when* $c=m$. (ii) *The absolute moment* $\mathbf{E}\{|x-c|\}$ *about a number* c *has the same value (called the mean deviation) for all median values* c *and a greater value when* c *is not a median.*

To prove (i), we observe that

$$\mathbf{E}\{(x-c)^2\}=\int(x-c)^2\,\mathrm{d}F=\int(x-m+m-c)^2\,\mathrm{d}F$$

$$=\int(x-m)^2\,\mathrm{d}F+2(m-c)\int(x-m)\,\mathrm{d}F+(m-c)^2\int\mathrm{d}F$$

$$=\sigma^2+(m-c)^2,$$

and the conclusion is obvious.

In (ii), if $c<q$, and q is a median,

$$\mathbf{E}\{|x-c|\}-\mathbf{E}\{|x-q|\}=\int(|x-c|-|x-q|)\,\mathrm{d}F$$

$$=q-c-2\int_c^q F(x)\,\mathrm{d}x$$

by partial integration. This vanishes if c is also a median value, since then $F(x)=\frac{1}{2}$ in $c<x<q$. But if c is not a median value, $F(x)<\frac{1}{2}$ in a proper subinterval of $c<x<q$,

$$\int_c^q F(x)\,\mathrm{d}x<\tfrac{1}{2}(q-c), \qquad \text{and} \qquad \mathbf{E}\{|x-c|\}>\mathbf{E}\{|x-q|\}.$$

A similar argument applies when $c>q$.

It is useful at this point to define some of the more familiar distributions and their properties.

The degenerate distribution

In the limiting case in which a variable takes a value λ with probability 1,

F has a single discontinuity at λ and is defined by

$$F(x)=0\ (x<\lambda), \qquad F(x)=1\ (x>\lambda).$$

It follows immediately that $\phi(t)=1$, $m=\lambda$, $\sigma=0$.

The binomial distribution

If $0<p<1$, $q=1-p$ and n is a positive integer, a random integer is said to have a binomial distribution (n, p) if it takes the value v with probability

$$p_v=\binom{n}{v}p^v q^{n-v}(v=0, 1, 2, \ldots n), \qquad p_v=0(v>n).$$

Then $m=np$, $\sigma^2=npq$, $p(z)=(q+pz)^n$,

$$\phi(t)=\sum_{v=0}^{n}\binom{n}{v}p^v q^{n-v}e^{itv}=(q+pe^{it})^n.$$

The Poisson distribution

The random integer can take all non-negative integral values v with probabilities $p_v e^{-c}c^v/c!$, where c is a positive constant. It follows immediately that $m=c$, $\sigma^2=c$,

$$p(z)=e^{c(z-1)}, \qquad \phi(t)=\exp[c(e^{it}-1)].$$

The Poisson distribution function can be regarded as the limiting form of the binomial distribution function when $pn=c$ is constant and n is large. For then

$$p_v=\binom{n}{v}p^v q^{n-v}=\frac{n(n-1)\ldots(n-v+1)}{v!}\frac{c^v}{n^v}\left(1-\frac{c}{n}\right)^{-v}\left(1-\frac{c}{n}\right)^n$$

$$=\frac{e^{-c}c^v}{v!}+o(1)$$

as $n\to\infty$ for each fixed v.

These three examples are all of discrete distributions. The four that follow are absolutely continuous and can be defined by their probability densities $F'(x)=f(x)$.

The rectangular distribution

The probability density of the **rectangular** or **uniform distribution** on an interval (a, b) is defined by

$$f(x)=(b-a)^{-1}(a\leqslant x\leqslant b), \qquad f(x)=0 \text{ elsewhere.}$$

Then plainly $m = q = \frac{1}{2}(a+b)$, $\sigma^2 = (b-a)^3/12$

$$\phi(t) = \frac{1}{b-a}\int_a^b e^{itx}\,dt = \frac{e^{itb} - e^{ita}}{it(b-a)}.$$

The normal (or Gaussian) distribution

We say that the distribution of a random real variable is **normal** (m, σ) if its probability density is given by

$$f(x) = (2\pi\sigma^2)^{-1/2}e^{-(x-m)^2/2\sigma^2}.$$

The fact that $\int f(x)\,dx = 1$ follows from the elementary identity

$$2\int_0^\infty e^{-x^2}\,dx = \pi^{1/2}.$$

It is plain that the median is m and it is easy to verify that the mean is m and the standard deviation σ as the notation suggests and anticipates, for

$$\mathbf{E}\{x\} = (2\pi\sigma^2)^{-1/2}\int xe^{-(x-m)^2/2\sigma^2}\,dx$$

$$= (2\pi)^{-1/3}\int (x+m)e^{-x^2/2} = m,$$

since $x\ \exp(-x^2/2)$ is an odd function. Further,

$$\mathbf{E}\{(x-m)^2\} = (2\pi\sigma^2)^{-1/2}\int (x-m)^2e^{-(x-m)^2/2\sigma^2}\,dx$$

$$= \sigma^2(2\pi)^{-1/2}\int x^2e^{-x^2/2}\,dx = \sigma^2$$

by partial integration.

The characteristic function is given by

$$\phi(t) = (2\pi\sigma^2)^{-1/2}\int e^{-(x-m)^2/2\sigma^2+itx}\,dx$$

$$= e^{itm-t^2\sigma^2/2}(2\pi)^{-1/2}\int e^{-(x+it\sigma)^2}\,dx,$$

and a simple contour integration shows that the integral is equal to

$$\int e^{-x^2/2}\,dx = (2\pi)^{1/2},$$

and so

$$\phi(t) = e^{itm-t^2\sigma^2/2}.$$

The Cauchy distribution

The density distribution given by $f(x)=a[\pi(x^2+a^2)]^{-1}$ with $a>0$ has median 0 but no mean and no finite standard deviation since $xf(x)$, $x^2f(x)$ are not integrable. It has characteristic function defined by

$$\phi(t)=\frac{a}{\pi}\int\frac{e^{itx}}{x^2+a^2}=\frac{1}{\pi}\int\frac{e^{itax}}{x^2+1}=e^{-a|t|},$$

again by a simple contour integration.

The χ^2 distribution

The χ^2 distribution of order k, for any positive integer k, is of great practical importance in statistics and is defined by the density

$$\begin{aligned} f_k(x) &= 2^{-k/2}[\Gamma(k/2)]^{-1}e^{-x/2}x^{k/2-1} && (x>0),\\ &= 0 && (x\leqslant 0). \end{aligned}$$

Its characteristic function, by contour integration, is given by

$$\phi_k(t)=[\phi_1(t)]^k=(1-2it)^{-k/2},$$

and it has mean $m=k$ and variance $\sigma^2=2k$.

6.3 Random real vectors

A random real vector $x=(x_1, x_2, \ldots, x_k)$ is defined by a probability measure μ in $\mathcal{R}^k$ which determines the joint distributions of its components $x_1, x_2, \ldots, x_k$. These are independent or not according as μ is or is not the product measure of k probability measures $\mu_1, \mu_2, \ldots, \mu_k$ in $\mathcal{R}$. There is no satisfactory analogue in $\mathcal{R}^k$ of the distribution function, which plays so useful a role in the one-dimensional case. However, there are two extreme types of the kind mentioned in the last section and which still merit a distinctive notation when they occur. First, we have a **point distribution** if a random vector can take only values λ_υ of a countable set and positive probabilities $p_\upsilon=\mathbf{P}\{x=\lambda_\upsilon\}$, with $\sum p_\upsilon=1$, are assigned. In particular, if the components of λ_υ all have non-negative integral values, the distribution can be defined by its **probability generating function** taking values

$$p(\xi)=\sum_{\upsilon_i\geqslant 0} p_\upsilon \xi_1^{\upsilon_1}\xi_2^{\upsilon_2}\ldots\xi_k^{\upsilon_k}.$$

At the other extreme, if μ is absolutely continuous, it has a non-negative derivative f by the Radon–Nikodym theorem such that

$$\mathbf{P}\{x\in X\}=\int_X f(x)\,dx=\int_X\ldots\int f(x_1, x_2, \ldots, x_k)\,dx_1, \ldots, dx_k$$

for every Borel measurable set X. The function f is naturally called the **probability density function** of x.

Many of the ideas and results of the last section can now be extended in a natural way to the k-dimensional case. The **mean** of a random vector x is the vector $m = (m_1, m_2, \ldots, m_k)$ defined by

$$m = \mathbf{E}\{x\} = \int x \, \mathrm{d}\mu, \qquad m_j = \int x_j \, \mathrm{d}\mu$$

in the general case and takes the form $m = \sum \lambda_v p_v$ for a point distribution. More generally, if α is any measurable vector valued function of x (not necessarily of the same dimension as x), its mean is given by

$$\mathbf{E}\{\alpha(x)\} = \int \alpha(x) \, \mathrm{d}\mu$$

whenever this integral exists. If the components of x are independent, this takes the form

$$\mathbf{E}\{\alpha(x)\} = \int \ldots \int \alpha(x) \, \mathrm{d}\mu_1 \ldots \mathrm{d}\mu_k.$$

The special cases $\alpha(x) = \sum x_j$ and $\alpha(x) = \prod x_j$ immediately give.

Theorem 10. (i) *If the random real vector* $x = (x_1, \ldots, x_k)$ *has mean* $m = (m_1, \ldots, m_k)$, *then* $\sum x_j$ *has mean* $\sum m_j$ (*whether* x_j *are independent or not*).

(ii) *If* x_j *are independent, then* $\prod x_j$ *has mean* $\prod m_j$.

If x has a probability density f, we have

$$\mathbf{E}\{\alpha(x)\} = \int \ldots \int \alpha(x) f(x) \, \mathrm{d}x_1 \ldots \mathrm{d}x_k,$$

and if the components are also independent, then $f(x) = f_1(x_1) \ldots f_k(x_k)$, where $f_1, f_2, \ldots, f_k$ are density functions over $\mathcal{R}$, and

$$\mathbf{E}\{\alpha(x)\} = \int \ldots \int \alpha(x) f_1(x_1) \ldots f(x_k) \, \mathrm{d}x_1 \ldots \mathrm{d}x_k.$$

The moments of x about its mean can be treated in the same way, the particularly important second moments being defined by

$$\mu_{ij} = \mathbf{E}\{(x_i - m_i)(x_j - m_j)\}, \qquad \sigma_i^2 = \mu_{ii} = \mathbf{E}\{X_i - m_i)^2\}.$$

The matrix $M = [\mu_{ij}]$ is called the **moment matrix**, its diagonal term μ_{ii} is called the **variance** of x_i, and σ_i is called the **standard deviation** of x_i. The ratio $\rho_{ij} = \mu_{ij}(\mu_{ii}\mu_{jj})^{-1/2}$ is called the **correlation coefficient** of x_i and x_j provided that $\mu_{ii} > 0$, $\mu_{jj} > 0$, and the matrix $[\rho_{ij}]$ which is defined when μ_{ii}

are all positive, is called the **correlation matrix**. The components x_i are said to be uncorrelated if $\mu_{ij} = 0$ when $i \neq j$, so that the moment matrix reduces to diagonal form. The correlation matrix then reduces to the unit matrix if the variances are all positive.

The **characteristic function** ϕ of a random vector x is defined by

$$\phi(t) = \phi(t_1, \ldots, t_k) = \mathbf{E}\{e^{it'x}\} = \int e^{it'x} \, d\mu,$$

where $t'x = \sum t_i x_i$ is the scalar product of t and x. The straightforward extension of Theorem 26 of Chapter 5 to vectors shows that a probability distribution is determined uniquely by its characteristic function. The characteristic function of a variable with a point distribution has values $\sum p_v e^{it'\lambda_v}$, which becomes $p[e^{it_1}, e^{it_2}, \ldots, e^{it_k}]$ when its components are non-negative integers.

Each of the variables x_i has its own characteristic function ϕ_i given by

$$\phi_i(t_i) = \mathbf{E}\{e^{it_i x_i}\} = \int e^{it_i x_i} \, d\mu,$$

and the relationship between these and the joint characteristic function for the vector $\boldsymbol{x}$ gives the following important criterion for independence.

Theorem 11. *If variables $x_1, x_2, \ldots, x_k$ have characteristic functions $\phi_1, \ldots, \phi_k$ and a joint characteristic function ϕ, a necessary and sufficient condition for independence is that*

$$\phi(t) = \phi_1(t_1)\phi_2(t_2) \ldots \phi_k(t_k).$$

The proof is almost immediate from the definition of independence, for if $x_1, x_2, \ldots, x_k$ are independent,

$$\phi(t) = \int e^{it'x} \, d\mu = \int \ldots \int e^{it_1 x_1} e^{it_2 x_2} \ldots e^{it_k x_k} \, d\mu_1 \ldots d\mu_k$$
$$= \phi_1(t_1) \ldots \phi_k(t_k),$$

since

$$\phi_j(t_j) = \int \ldots \int e^{it_j x_j} \, d\mu_1 \ldots d\mu_k = \int e^{it_j x_j} \, d\mu_j.$$

Conversely, if $\phi(t) = \phi_1(t_1) \ldots \phi_k(t_k)$, we have

$$\phi(t) = \int e^{it_1 x_1} \, d\mu_1 \ldots \int e^{it_k x_k} \, d\mu_k = \int e^{it'x} \, d\mu,$$

where μ is the product measure of $\mu_1, \mu_2, \ldots, \mu_k$ and is the distribution of a vector with independent components.

If $\boldsymbol{x}$ has independent components with the same distribution μ in $\mathcal{R}$, it is usually called a **sample** of size k from a population with distribution μ, and the study of samples and distributions of functions of samples is of great practical importance.

In the more general case, we are usually concerned with a vector valued function of a random vector. In some cases it is enough to know its mean and other parameters, while in others we require its complete distribution. We devote the rest of this section to some general theorems and special examples relating to these ideas.

We begin with the simplest algebraic function of a vector, defined by its linear transform $y = Cx$ by a matrix C, and obtain the following generalization of Theorem 10(i).

Theorem 12. *If x is a random vector in $\mathcal{R}^k$ with mean m and characteristic function ϕ, C a matrix of l rows and k columns, the random l-vector defined by $y = Cx$ has mean Cm and characteristic function $\omega(u) = \phi(C'u)$.*

If μ denotes the distribution of x in $\mathcal{R}^k$, we have

$$\mathbf{E}\{y_i\} = \mathbf{E}\left\{\sum_{j=1}^{k} c_{ij}x_j\right\} = \int \sum_j c_{ij}x_j \, d\mu = \sum_j c_{ij} \int x_j \, d\mu$$

$$= \sum c_{ij}\mathbf{E}\{x_j\} = \sum c_{ij}m_j,$$

and since this holds for $i = 1, 2, \ldots, l$, we get $\mathbf{E}\{y\} = Cm$. Also

$$\psi(u) = \mathbf{E}\{e^{iu'y}\} = \mathbf{E}\{e^{iu'Cx}\} = \mathbf{E}\{e^{i(C'u)'x}\} = \phi(C'u)$$

When the terms of x are independent, we have the following important result.

Theorem 13. *If $x_1, x_2, \ldots, x_k$ are independent real variables with distribution functions $F_1, F_2, \ldots, F_k$ and characteristic functions $\phi_1, \ldots, \phi_k$, respectively, their sum has distribution function $F = F_1 * dF_2 * \ldots * dF_k$ and characteristic function $\psi = \phi_1\phi_2 \ldots \phi_k$. Moreover, if at least one variable has a density distribution, so has their sum.*

Here the characteristic function ϕ of the vector x in $\mathcal{R}^k$ is defined by

$$\phi(t) = \phi_1(t_1) \ldots \phi_k(t_k)$$

by Theorem 11, and if C is the row vector $(1, 1, \ldots, 1)$, the last theorem gives

$$y = x_1 + x_2 + \ldots + x_k = Cx,$$

$$\psi(u) = \phi(C'u) = \phi(u, u, \ldots, u) = \phi_1(u) \ldots \phi_k(u).$$

The results about the distribution functions then follow from Theorem 33 of Section 5.5.

There is an analogous but more elementary theorem for random integers.

Theorem 14. *x_1, x_2 are random non-negative integers with probability generating functions p_1, p_2, then x_1+x_2 has generating function p_1p_2.*

We need only observe that

$$p_v = \mathbf{P}\{x_1+x_2=v\} = \sum_{j=0}^{v} \mathbf{P}\{x_1=j, x_2=v-j\}$$
$$= \sum_{j=0}^{v} p_{1,j}p_{2,v-j},$$

so that

$$p(\xi) = \sum_{v=0}^{\infty} p_v\xi^v = p_1(\xi)p_2(\xi).$$

The following theorem shows that sums of pairs of random variable of certain types are also of the same types.

Theorem 15. *We suppose that x_1, x_2 are independent. Then*

(i) *if x_1, x_2 are binomial (n_1, p), (n_2, p), respectively, then x_1+x_2 is binomial (n_1+n_2, p).*
(ii) *if x_1, x_2 are both Poisson variables with means c_1, c_2, then x_1+x_2 is Poisson with mean c_1+c_2.*
(iii) *if x_1, x_2 are normal (m_1, σ_1) and (m_2, σ_2), respectively, then x_1+x_2 is normal (m, σ), where $m=m_1+m_2$, $\sigma^2=\sigma_1^2+\sigma_2^2$.*
(iv) *if x_1, x_2 are $\chi^2(l_1)$ and $\chi^2(l_2)$, respectively, then x_1+x_2* is *$\chi^2(l_1+l_2)$.*

These follow at once from Theorem 13 and the forms of the characteristic functions for the special distributions listed in Section 6.2.

The extension to quadratic functions of a random vector can be based very conveniently on the moment matrix.

Theorem 16. *Suppose that x is a random vector in $\mathcal{R}^k$ with moment matrix M. Then $y=Cx$ has moment matrix CMC'.*

If we suppose that $\boldsymbol{x}$ has mean m so that $q=Cm$ is the mean of $\mathbf{y}$, the second-order moments v_{ij} of $\mathbf{y}$ are given by

$$v_{ij} = \mathbf{E}\{(y_i-q_i)(y_j-q_j)\} = \mathbf{E}\left\{\sum_{r=1}^{k}\sum_{s=1}^{k} C_{ir}C_{js}(x_r-m_r)(x_s-m_s)\right\}$$
$$= \sum_{r=1}^{k}\sum_{s=1}^{k} C_{ir}C_{js}\mathbf{E}\{(x_r-m_r)(x_s-m_s)\} \qquad \text{(by Theorem 11)}$$
$$= \sum_{r=1}^{k}\sum_{s=1}^{k} C_{ir}C_{js}\mu_{rs},$$

which is our conclusion.

As an immediate corollary, we have

Theorem 17. *If $x_1, x_2, \ldots, x_k$ are uncorrelated (in particular, if they are independent) and have variances $\sigma_1^2, \sigma_2^2, \ldots, \sigma_k^2$, then their sum has variance $\sigma_1^2+\sigma_2^2+\ldots+\sigma_k^2$.*

In this case M is diagonal, with diagonal terms σ_j^2, C is the row vector with unit terms and the variance of $y=\sum x_j = Cx$ is the single term of the 1×1 matrix CMC', which is plainly $\sum \sigma_j^2$.

Theorem 18. *If x is a sample of size k from a population of mean m and variance σ^2, the sample mean $\bar{x}=k^{-1}\sum x_j$ has mean m and variance σ^2/k, and the sample variance $s^2=k^{-1}\sum (x_j-\bar{x})^2$ has mean $(k-1)\sigma^2/k$.*

The mean and variance of $\bar{x}$ are given at once by Theorems 12 and 17. Using Theorem 12 again, we have

$$\mathbf{E}\{s^2\}=\frac{1}{k}\sum_{j=1}^{k}\mathbf{E}\{(x_j-\bar{x})^2\}.$$

It is sufficient to take the case $m=0$, and then

$$x_1-\bar{x}=\frac{1}{k}[(k-1)x_1-x_2-\ldots-x_k]$$

and

$$\mathbf{E}\{(x_1-\bar{x})^2\}=\frac{\sigma^2}{k^2}[(k-1)^2+(k-1)]=(k-1)\sigma^2/k,$$

since $\mathbf{E}\{x_ix_j\}=0$ when $i\neq j$ and $\mathbf{E}\{x_j^2\}=\sigma^2$. It is plain that $\mathbf{E}\{(x_j-\bar{x})^2\}$ has the same value for $j=1,2,\ldots,k$, and therefore $\mathbf{E}\{s^2\}=(k-1)\sigma^2/k$, as required.

We now describe briefly some of the more familiar special distributions in $\mathcal{R}^k$.

The multinomial distribution $(n: p_1, p_2, \ldots, p_k)$

We suppose that $q>0$, $p_j>0$, $\sum p_j=1-q$, that n is a positive integer and that the components $(x_1, x_2, \ldots, x_k)$ of a random vector take non-negative integral values with probabilities defined by

$$\begin{aligned}\mathbf{P}\{x_j=v_j, j=1,2,\ldots,k\}&\\ &=\frac{n!}{v_1!\,v_2!\ldots v_k!\,u!}p_1^{v_1}p_2^{v_2}\ldots p_k^{v_k}q^u, \quad \text{if} \quad u=n-\sum_j v_j\geqslant 0\\ &=0 \quad \text{if} \quad u<0.\end{aligned}$$

The binomial distribution in $\mathcal{R}$ which was defined in the last section is obviously the case $k=1$, $p_1=p$, $q=1-p$. The distribution in the general case is in $\mathscr{R}^k$, but it may equally well be regarded, if we write $u = v_{k+1}$, $q = p_{k+1}$, as a distribution over the hyperplane

$$\sum_{j=1}^{k+1} x_j = n \qquad \text{in } \mathscr{R}^{k+1}.$$

The probability generating function p is given by

$$p(\xi) = \sum_{v_1+v_2+\ldots+v_k \leqslant n} \frac{n!\, q^u}{v_1!\, v_2! \ldots v_k!\, u!} (p_1\xi_1)^{v_1} \ldots (p_k\xi_k)^{v_k}$$
$$= (q + p_1\xi_1 + p_2\xi_2 + \ldots + p_k\xi_k)^n,$$

and the mean $m = (m_1, m_2, \ldots, m_k)$ by $m_j = \partial p/\partial\xi_j$ at the point $\xi = (1, 1, \ldots, 1)$, and this is np_j.

The characteristic function ϕ is given by

$$\phi(t) = p(e^{it_1}, e^{it_2}, \ldots, e^{it_k}) = \left(q + \sum_{j=1}^{k} p_j e^{it_j}\right)^n.$$

The k-dimensional Poisson distribution

We suppose that $c_j > 0$ $(j = 1, 2, \ldots, k)$ and that $x_1, x_2, \ldots, x_k$ take non-negative integral values with probabilities given by

$$p\{x_j = v_j, j = 1, 2, \ldots, k\} = e^{-c} \frac{c_1^{v_1} c_2^{v_2} \ldots c_k^{v_k}}{v_1!\, v_2! \ldots v_k!}\Bigg),$$

where $c = \sum c_j$. The probability generating function is given by

$$p(\xi) = e^{-c} \sum_{v_j \geqslant 0} \frac{(c_1\xi_1)^{v_1}(c_2\xi_2)^{v_2} \ldots (c_k\xi_k)^{v_k}}{v_1!\, v_2! \ldots v_k!}$$
$$= \prod_{j=1}^{k} e^{c_j(\xi_j - 1)}.$$

The mean is defined by

$$m_j = \left(\frac{\partial p}{\partial \xi_j}\right)_{\xi = 1,1,\ldots,1} = c_j,$$

and the characteristic function ϕ is given by

$$\phi(t) = p(e^{it_1}, e^{it_2}, \ldots, e^{it_k}) = \prod_{j=1}^{k} \exp[c_j(e^{it_j} - 1)].$$

The uniform distribution over a set

If X is a measurable set of Lebesgue measure A in $\mathscr{R}^k$, the distribution with constant density $f(x) = A^{-1}$ in X and $f(x) = 0$ in X' defines a uniform distribution over X. If $k = 1$, the most natural form of X is an interval, and we get the rectangular distribution defined before. If $k \geq 2$, other forms such as circles or rectangles may occur. For example, the uniform distribution over the rectangle $|x_1| \leq a_1$, $|x_2| \leq a_2$ in $\mathcal{R}^2$ has mean 0, density function $(4a_1a_2)^{-1}$ in the rectangle and characteristic function

$$\left(\frac{\sin a_1t_1}{a_1t_1}\right)\left(\frac{\sin a_2t}{a_2t}\right).$$

The normal distribution

We begin by defining an important special case. We say that a vector x in $\mathcal{R}^k$ has an independent normal distribution, with mean m and variances $\sigma_1^2, \sigma_2^2, \ldots, \sigma_k^2$, if its components x_j are independent and normal and have means m_j, respectively. The density of x is then plainly

$$(2\pi)^{-n/2}(\sigma_1\sigma_2\ldots\sigma_k)^{-1/2}\exp[-\sum (x_j - m_j)^2/2\sigma_j^2],$$

and its characteristic function has value $\exp(it'm - \frac{1}{2}\sum t_j^2\sigma_j^2)$.

We now define a general normal distribution by saying that x is a normal vector in $\mathcal{R}^k$ if it can be expressed by $x = m + Cy$, where m is a constant vector in $\mathcal{R}^k$, y is an independent normal vector in $\mathcal{R}^l$ and C is a matrix of k rows and l columns. It is obvious from the definition that any linear transform of a normal vector is also normal, and the other essential properties of normal vectors and distributions are contained in the following theorems.

Theorem 19. *The normal variable $x = m + Cy$, where y is independent in $\mathcal{R}^l$ with mean 0 and variances $\sigma_1^2, \sigma_2^2, \ldots, \sigma_l^2$ and C has rank r, has mean m and characteristic function defined by*

$$\phi(t) = e^{it'm - (1/2)t'Mt},$$

where $M = CC'$ is a non-negative $k \times k$ matrix of rank r.

We deduce immediately from Theorem 12 that the mean of x is m and that

$$\phi(t) = e^{it'm}\psi(C't),$$

where $\psi(u)$ is the characteristic function of $\mathbf{y}$ and has the form

$$\psi(u) = \exp(-\tfrac{1}{2}\sum u_j^2\sigma_j^2) = e^{-(1/2)u'\mathbf{Q}u}$$

and $\mathbf{Q}$ has terms $\sigma_1^2, \sigma_2^2, \ldots, \sigma_l^2$ in its first l diagonal places and zeros elsewhere. We therefore have the required formula

$$\phi(t) = e^{it'm - (1/2)t'Mt} \qquad \text{with} \qquad M = CQC'.$$

It is obvious from the form of M that it is symmetric, non-negative and of order k.

Conversely,

Theorem 20. *If M is a non-negative symmetric $k \times k$ matrix of rank r, then*

$$\phi(t) = e^{it'm-(1/2)t'Mt}$$

defines the characteristic function of a normal random variable x.

In every expression of x in the form $x = m + Cy$ in which y is an independent normal vector, the matrix C has the same rank r. Moreover, an expression of this form can be found in which $l = r$, the variances $\sigma_1^2, \sigma_2^2, \ldots, \sigma_r^2$ of y are the r positive proper values of M and C consists of the first r columns of an orthogonal matrix P.

Alternatively, we can take $x = m + P\eta$, where η is a vector in $\mathcal{R}^k$ whose first r components are independent and normal $(0, \sigma_j)$ $(j = 1, 2, \ldots, r)$ and whose remaining $k - r$ components are all zero.

The **rank** of **a normal vector** is defined as the common rank of all the matrices C which may be used in its definition. The vector and its distribution are called **regular** if $r = k$, $|M| > 0$ and **singular** if $r < k$, $|M| = 0$. We have as immediate corollaries.

Theorem 21. *The rank of a normal vector is the smallest number of independent normal variables in terms of which it can be expressed linearly.*

Theorem 22. *A normal vector of rank r lies with probability 1 in a sub-space of dimension r.*

Theorem 23. *A normal random vector x has a density distribution if and only if it is regular. In this case, if its characteristic function is given by*

$$\phi(t) = e^{it'm-(1/2)t'Mt},$$

its density function by

$$f(x) = (2\pi)^{-n/2} |M|^{-1/2} \exp[-\tfrac{1}{2}(x-m)'M^{-1}(x-m)],$$

where M is the moment matrix of x, $\sigma_1^2, \sigma_2^2, \ldots, \sigma_k^2$ its proper values and

$$|M| = \sigma_1^2\sigma_2^2 \ldots \sigma_k^2.$$

Furthermore, we can write $x = m + Py$, where P is orthogonal and y an independent normal vector in $\mathcal{R}^k$ with mean 0 and variances σ_j^2.

By a familiar theorem on matrices, we can define an orthogonal matrix P of order k so that $M = P'QP$. Then

$$\phi(t) = e^{it'm-(1/2)t'Mt} = e^{it'm-(1/2)t'P'QPt}$$

and this, after Theorem 11, is the characteristic function of $m + Cy = m + P\eta$, where y_j are independent and normal $(0, \sigma_j)$ for $1 \leq j \leq r$ and $\eta_j = y_j$ for $1 \leq j \leq r$, $\eta_j = 0$ for $r + 1 \leq j \leq k$.

The joint distribution of the mean and variance of a normal sample

It has been shown in Theorem 18 that the mean and standard deviation of the mean $\bar{x}$ and the variance s^2 of a sample of size k from any given population can be found in terms of those of the population. It is not generally possible to go further and find the distributions of $\bar{x}$ and s, either separately or jointly, except for special populations. Among these, the normal population is particularly important in practical applications, and the next theorem gives the complete joint distribution of $\bar{x}$ and s^2 and shows that they are independent.

Theorem 24. *If $x=(x_1, x_2, \ldots, x_k)$ is a sample from a normal $(0, 1)$ population, then the sample mean $\bar{x}$ and variance s^2 are independent. The mean $\bar{x}$ is normal $(0, k)^{-1/2})$ and ks^2 has the χ^2 distribution of order $k-1$ with density $f_{k-1}(x)$ and characteristic function ψ_{k-1}, where*

$$
\begin{aligned}
f_k(x) &= e^{-(1/2)x}x^{(1/2)k-1} \qquad && (x>0),\\
&= 0 && (x\leqslant 0),\\
\psi_k(u) &= (1-2iu)^{-k/2} = [\psi_1(u)]^k.
\end{aligned}
$$

We can define an orthogonal transformation $x=Py$ by means of an orthogonal matrix P so that $y_1, y_2, \ldots, y_k$ are independent, $y_k = k^{-1/2}\sum x_j = \bar{x}k^{1/2}$. Then

$$
ks^2 = \sum_{j=1}^{k} (x_j-\bar{x})^2 = \sum_{j=1}^{k} x_j^2 - k\bar{x}^2 = \sum_{j=1}^{k} y_j^2 - y_k^2,
$$

by the orthogonality property of P, and so

$$
ks^2 = \sum_{j=1}^{k-1} y_j^2,
$$

and we get the joint characteristic function $\phi(t, u)$ of $(\bar{x}, ks^2)$ in the form

$$
\begin{aligned}
\phi(t, u) &= \mathbf{E}(\{\exp(it\bar{x}+iuks^2)\} = \mathbf{E}\{\exp(itk^{-1/2}y_k + iu\sum y_j^2)\}\\
&= \mathbf{E}\{\exp(itk^{-1/2}y)\}\prod_{j=1}^{k-1}\mathbf{E}\{\exp(iuy_j^2)\}
\end{aligned}
$$

by Theorem 12, since $y_1, y_2, \ldots, y_k$ are independent. Hence

$$
\phi(t, u) = \exp(-t^2/2k)[\psi_1(u)]^{k-1},
$$

where

$$
\begin{aligned}
\psi_1(u) &= \mathbf{E}\{\exp(iuy^2)\} = (2\pi)^{-1/2}\int \exp[(iu-\tfrac{1}{2})y^2]\,dy\\
&= (2\pi)^{-1/2}\int_0^\infty e^{-1/2x(1-2iu)}x^{-1/2}\,dx = \int_0^\infty e^{iux}f_1(x)\,dx,
\end{aligned}
$$

and this is $(1-2iu)^{-1}$ by simple contour integration. Since ψ_1 is the characteristic function of the density $f_1(x)$, the formula for $f_{k-1}(x)$ follows from the relation $f_k = f_{k-1} * f$, which is obvious by induction. The independence of $\bar{x}$ and s^2 is clear from the factorization of $\phi(t, u)$ into their characteristic functions $\exp(-t^2/2k)$ and ψ_{k-1}.

6.4 Dependence and conditional probability

The idea of conditional probability can be introduced most easily by considering the product space $\mathscr{Z} = \mathscr{X} \times \mathscr{Y}$ of spaces $\mathscr{X}$, $\mathscr{Y}$ and supposing that a joint probability distribution λ is given for the random variable (x, y). We know after Section 2.5· that this joint distribution defines probability distributions $\mu(X) = \lambda(X \times \mathscr{Y})$, $\nu(Y) = \lambda(\mathscr{X} \times Y)$ in $\mathscr{X}$, $\mathscr{Y}$, respectively, but do not suppose that these are independent.

If X, Y are measurable sets in $\mathscr{X}$, $\mathscr{Y}$, respectively, we know that $X \times Y$ is measurable in $\mathscr{Z}$ and if $\mu(X) > 0$, we call $\lambda(X \times Y)/\mu(X)$ the **conditional probability** that y belongs to Y under the condition that x belongs to X. We denote this by $\nu(Y/X)$, so that

$$\nu(Y/X) = \frac{\lambda(X \times Y)}{\mu(X)} \qquad (\mu(X) > 0).$$

It is clear that, for any fixed X, $\nu(Y/X)$ is defined in the σ-ring of measurable sets Y of $\mathscr{Y}$, and is completely additive by the complete additivity of λ. Moreover, $\nu(Y/X) \geqslant 0$ and

$$\nu(\mathscr{Y}/X) = \frac{\lambda(X \times \mathscr{Y})}{\mu(X)} = 1,$$

so that $\nu(Y/X)$ is a probability measure in $\mathscr{Y}$ for each fixed X. In particular, if X reduces to a single point x with $\mu(X) > 0$, we write $\nu(Y/X)$ as $\nu(Y/x)$. By interchanging X and Y in these arguments, we have

$$\mu(X/Y) = \frac{\lambda(X \times Y)}{\nu(Y)} \qquad (\nu(Y) > 0)$$

as the conditional probability that $x \in X$ under condition $y \in Y$, and there is complete symmetry between X and Y. In application, however, it is generally more useful to think of y as being in some sense dependent on x, and the extreme case of this arises when the value of y is strictly determined (or determined with probability 1) by that of x, and so that y is the value at x of a function in the usual real variable sense and the whole mass of the joint distribution is concentrated in the graph $[x, y(x)]$. The values, or sets of values, of x can now be regarded as **hypotheses,** the values of y as **observations**. A conditional probability then gives the expected distribution of the observed variable under a given hypotheses, and experiments can often be designed so that these conditional probabilities are known for each of a system of possible hypotheses. The object of the experiment is then to use the observations of values of x to express opinions about the merits of different hypotheses or to modify any opinions which were held before the experiment took place. For this reason, the distribution μ of x is usually called the **prior probability** distribution of the hypothesis x and the conditional probability $\mu(X/Y)$ is called the **post probability** or **likelihood** of X under the observation Y.

These ideas are illustrated in the following famous result called **Bayes' Theorem**.

Theorem 25. *Suppose that* (x, y) *is a random variable in the product space* $\mathscr{X}\times\mathscr{Y}$, *that* Y, X_k *are measurable and disjoint for* $k=1,2,3,\ldots$ *and* $\nu(Y)>0$, $\mu(X_k)>0$, $\mathscr{X}=\bigcup_{k=1}^{\infty} X_k$. *Then*

$$\mu(X_j/Y)=\frac{\nu(Y/X_j)\mu(X_j)}{\sum \nu(Y/X_k)\mu(X_k)}.$$

This follows almost immediately from the definitions, since

$$\mu(X_j/Y)\sum \nu(Y/X_k)\mu(X_k)=\frac{\lambda(X_j\times Y)}{\nu(Y)}\sum \lambda(X_k\times Y)$$

$$=\frac{\lambda(X_j\times Y)}{\nu(Y)}\lambda(\mathscr{H}\times Y)=\lambda(X_j\times Y)=\nu(Y/X_j)\mu(X_j).$$

The concept of conditional probability can be put in a way which appears to be more general. If $\mathscr{Z}$ is any probability space with distribution λ, and Z_1 any measurable subset of it, we can define the conditional probability $\lambda(Z/Z_1)$ that $z\in Z$ under the condition $z\in Z_1$ to be $\lambda(Z\cap Z_1)/\lambda(Z_1)$. The definitions used above then apply if the sets Z, Z_1 are restricted to sets of the type $X\times Y$ in a product space $\mathscr{Z}=\mathscr{X}\times\mathscr{Y}$. Bayes' Theorem takes the form

$$\lambda(Z_k/Z)=\frac{\lambda(Z/Z_k)\lambda(Z_k)}{\sum \lambda(Z/Z_k)\lambda(Z_k)}$$

where Z_k are measurable sets and $\lambda(Z_k)>0$, $\bigcup Z_k=\mathscr{Z}$. In fact, there is no greater generality in this form since it can be put in the form of Theorem 25 by considering the product space $\mathscr{Z}\times\mathscr{X}$, where $\mathscr{X}$ is the space of positive integers with $\mathbf{P}\{x=k\}$ defined to be $\lambda(Z_k)$.

We shall therefore maintain the product space form as it was first introduced and illustrate the application of Bayes' Theorem to a simple experiment.

We suppose that two boxes x_1, x_2 are offered at random with (prior) probabilities $\frac{1}{2}$, $\frac{2}{3}$. It is known that the box x_1 contains 8 white counters and 12 red, while x_2 contains 4 white and 4 red counters. The observer takes a counter at random from the box offered, without knowing which it is, and finds that it is white. How will he assess the likelihood that the box was x_1?

Here, the **hypothesis space** $\mathscr{X}$ contains two points x_1, x_2 with probabilities $\mu(x_1)=\frac{1}{3}$, $\mu(x_2)=\frac{2}{3}$. The **observation space** $\mathscr{Y}$ also contains only two values y_1 (for a white counter), y_2 (for red). The conditional probabilities are given by the proportions of red and white counters in each box, so that

$$\nu(y_1/x_1)=8/20=2/5, \qquad \nu(y_1/x_2)=4/8=1/2.$$

Hence, by Bayes' theorem, the likelihood that the box is x_1 is

$$\begin{aligned}\mu(x_1/y_1)&=\frac{\nu(y_1/x_1)\mu(x_1)}{\nu(y_1/x_1)\mu(x_1)+\nu(y_1/x_2)\mu(x_2)}\\ &=\frac{(2/5\times 1/3)}{(2/5)(1/3)+(1/2)(2/3)}=2/7.\end{aligned}$$

The result may be stated in terms of the frequency interpretation of probabilities by saying that in a long sequence of experiments, the proportion of offers of box x_1 in those experiments which resulted in the choice of a white counter would be 2/7.

The conditional probability $\nu(Y/x)$ has been defined for points x only when x has positive μ measure when considered as a measurable set, and it is natural to enquire whether any extension to more general cases is possible. This can be done, but rather less directly than before, through the following theorem.

Theorem 26. *Let λ be the joint distribution of (x, y) in $\mathscr{X}\times\mathscr{Y}$ and let μ, ν be the distribution of x, y, respectively. Then if Y is any measurable set with respect to ν, there is defined for almost all x the value $\nu(Y/x)$ of a unique function which is integrable over $\mathscr{X}$ with respect to μ and has the property that*

$$\lambda(X\times Y)=\int_X \nu(Y/x)\,\mathrm{d}\mu$$

for every measurable X in $\mathscr{X}$. In particular, $\nu(Y)=\lambda(\mathscr{X}\times Y)=\int_{\mathscr{X}}\nu(Y/x)\,\mathrm{d}\mu$, and $\nu(\mathscr{Y}/x)=1$ for almost all x.

If we consider $\lambda(X\times Y)$, for fixed Y, as the value of a set function defined on the σ-ring of measurable sets X, it is plain from the fact that $\lambda(Z)$ is a probability measure that it is completely additive. Moreover, since

$$\lambda(X \times Y) \leqslant \lambda(X \times \mathscr{Y}) = \mu(X),$$

it is absolutely continuous with respect to μ, and the main conclusion then follows from the Radon–Nikodym theorem (Theorem 28 of Chapter 2). That $\nu(\mathscr{Y}/x) = 1$ a.e. follows from the fact that $0 \leqslant \nu(\mathscr{Y}/x) \leqslant 1$ and $\mu(\mathscr{X}) = 1 = \lambda(\mathscr{X} \times \mathscr{Y}) = \int_{\mathscr{X}} \nu(\mathscr{Y}/x)\,\mathrm{d}\mu$.

The number $\nu(Y/x)$ is called the **conditional probability** that y belongs to Y under the condition that x takes its assigned value. It is defined uniquely outside a null set by λ, and its value is the same as that previously defined in the case when $\mu(x) > 0$. It is important to observe, however, that the null set in $\mathscr{X}$ in which $\nu(Y/x)$ is not defined depends on Y, so that although the definition of $\nu(Y/x)$ can be extended simultaneously to any *countable* system of sets Y by excluding the countable union of all the null sets associated with each, it cannot be done in general for non-countable systems. Above all, we cannot assume in general that, for almost all x, $\nu(Y/x)$ is completely additive in Y and has a measure extension. We shall return to this point later, but show at this stage that it is possible to proceed somewhat further even without any extra assumptions. First, we deduce another form of Bayes' Theorem.

Theorem 27. *If $\nu(Y) > 0$, the post probability $\mu(X/Y)$ under the observation Y has a density $\nu(Y/x)/\nu(Y)$ with respect to μ, and Bayes' formula can be written in the form*

$$\mu(X/Y) = \frac{\int_X \nu(Y/x)\,\mathrm{d}\mu}{\nu(Y)} = \frac{\int_X \nu(Y/x)\,\mathrm{d}\mu}{\int_{\mathscr{X}} \nu(Y/x)\,\mathrm{d}\mu}$$

for every measurable set X.

This follows immediately from Theorem 26 and the definition of $\nu(Y/X)$ and $\mu(X/Y)$, since

$$\mu(X/Y) = \frac{\lambda(X \times Y)}{\nu(Y)} = \frac{\lambda(X \times Y)}{\lambda(\mathscr{X} \times Y)}.$$

The conditional probability $\nu(Y/X)$ is a measure in $\mathscr{Y}$ and if a real-valued function α is integrable with respect to it, we have the **conditional mean** $\mathbf{E}\{\alpha(y)/X\}$ of $\alpha(y)$ given by

$$\mathbf{E}\{\alpha(y)/X\} = \int_{\mathscr{Y}} \alpha(y)\,\mathrm{d}\nu(Y/X).$$

The following theorem shows that a conditional mean $\mathbf{E}\{\alpha(y)/x\}$ can be defined for *points* x, in spite of the fact that $\nu(Y/x)$ is not generally a measure. When it is a measure, we see below that $\mathbf{E}\{\alpha(y)/x\}$ can be expressed as an integral in the normal way.

Theorem 28. *Suppose that α is integrable with respect to ν. Then it is integrable with respect to $\nu(Y/X)$ whenever X is measurable and $\mu(X)>0$. Moreover, there is defined for almost all x the value $\mathbf{E}\{\alpha(y)/x\}$ of a unique function which is integrable over $\mathscr{X}$ with respect to μ and has the property that*

$$\mathbf{E}\{\alpha(y)/X\}=\int_{\mathscr{Y}}\alpha(y)\,\mathrm{d}\nu(Y/X)=[\mu(X)]^{-1}\int_X \mathbf{E}\{\alpha(y)/x\}\,\mathrm{d}\mu$$

whenever $\mu(X)>0$. In particular, if $X=\mathscr{X}$,

$$\mathbf{E}\{\alpha(y)\}=\mathbf{E}\{\alpha(y)/\mathscr{X}\}=\int_{\mathscr{X}}\mathbf{E}\{\alpha(y)/x\}\,\mathrm{d}\mu.$$

If α is the characteristic function of a measurable set Y, then

$$\mathbf{E}\{\alpha(y)/X\}=\nu(Y/X),\qquad \mathbf{E}\{\alpha(y)/x\}=\nu(Y/x),$$

and the theorem reduces to Theorem 26.

If α is the characteristic function of a measurable set Y and $\mu(X)>0$, $\nu(Y)>0$, we have

$$\mu(X)\int_{\mathscr{Y}}\alpha(y)\,\mathrm{d}\nu(Y/X)=\mu(X)\nu(Y/X)=\lambda(X\times Y)=\int_{X\times\mathscr{Y}}\alpha(y)\,\mathrm{d}\lambda$$

and the same is obviously true for any simple function α. Also, we have

$$\mu(X)\int|\alpha(y)|\,\mathrm{d}\nu(Y/X)\leqslant\int|\alpha(y)|\,\mathrm{d}\nu$$

for every simple function α, since $\mu(X)\nu(Y/X)=\lambda(X\times Y)\leqslant\nu(Y)$ for every Y. This is enough to ensure that any function α which is integrable with respect to ν is also integrable over $\mathscr{Y}$ with respect to $\nu(Y/X)$ and over $\mathscr{Z}$ with respect to λ, and that

$$\mu(X)\mathbf{E}\{\alpha(y)/X\}=\mu(X)\int\alpha(y)\,\mathrm{d}\nu(Y/X)=\int_{X\times\mathscr{Y}}\alpha(y)\,\mathrm{d}\mu.$$

This is true, in particular, when $X=\mathscr{X}$, and therefore $\int_Z\alpha(y)\,\mathrm{d}\lambda$ is absolutely continuous and bounded over the σ-ring of measurable sets Z in $\mathscr{Z}$. It is then absolutely continuous over the σ-ring of sets of the form $X\times\mathscr{Y}$ and it follows from the Radon–Nikodym theorem that $\int_{X\times\mathscr{Y}}\alpha(y)\,\mathrm{d}\lambda$ is the integral over X of a function whose values $\mathbf{E}\{\alpha(y)/x\}$ are defined uniquely almost everywhere and which is integrable in $\mathscr{X}$. The conclusions of the theorem follow at once.

The next theorem deals with the case in which the conditional probabilities $\nu(Y/x)$ can be extended into probability measures and shows that the expressions $\mathbf{E}\{\alpha(y)/x\}$ can be expressed as integrals with respect to $\nu(Y/x)$ in the usual sense.

Theorem 29. *Suppose that for almost all x a probability measure $\nu_x(Y)$ can be defined on the measurable sets Y of $\mathcal{Y}$ to satisfy*

$$\lambda(X\times Y)=\int_X \nu_x(Y)\,d\mu$$

for every measurable X.

Then if Y is any measurable set, $\nu_x(Y)=\nu(Y/x)$ for almost all x, and if $\alpha(x, y)$ is integrable in $\mathcal{X}$, then $\alpha(x, y)$ is integrable in $\mathcal{Y}$ with respect to ν_x for almost all x and

$$\mathbf{E}\{\alpha(x, y)\}=\int \alpha(x, y)\,d\lambda=\int_{\mathcal{X}}\left\{\int_{\mathcal{Y}}\alpha(x, y)\,d\nu_x\,d\mu\right\}.$$

In particular, if $\alpha(y)$ is integrable with respect to ν, it is integrable with respect to ν_x and

$$\mathbf{E}\{\alpha(y)/x\}=\int \alpha(y)\,d\nu_x$$

for almost all x.

Moreover, the distribution is independent if and only if $\nu_x(Y)=\nu(Y)$, for almost all x.

The first part is obvious from the fact that

$$\int_X \nu(Y/x)\,d\mu=\lambda(X\times Y)=\int_X \nu_x(Y)\,d\mu$$

and

$$\int_X [\nu(Y/x)-\nu_x(Y)]\,d\mu=0$$

for every X. For the next part, we suppose first that α is the characteristic function of a set $X\times Y$, when X, Y are both measurable. Using Theorem 26, we have

$$\nu(Y/x)=\int_Y \alpha(x, y)\,d\nu_x,$$

and

$$\int_X \nu(Y/x)\,d\mu=\int\left\{\int\alpha(x, y)\,d\nu_x\right\}d\mu=\int\alpha(x, y)\,d\lambda=\lambda(X\times Y)$$

and this extends immediately to simple functions, and for such functions we have also

$$\int\left\{\int |\alpha(x, y)|\,d\nu_x\right\}d\mu \leqslant \int |\alpha(x, y)|\,d\lambda.$$

This is enough to ensure by the approximation to α by simple functions that $\int |\alpha(x, y)|\,d\nu_x$ exists for almost all x if α is a general integrable function over $\mathscr{Z}$ and there is then no difficulty in extending the formula

$$\int\left\{\int \alpha(x, y)\,d\nu_x\right\}d\mu = \int \alpha(x, y)\,d\lambda$$

from simple functions to the general case. The special form

$$\mathbf{E}\{\alpha(y)/x\} = \int \alpha(y)\,d\nu_x$$

follows by taking $\alpha(x, y) = \alpha(y)$.

Finally, if $\nu_x = \nu$ for almost all x, we have

$$\lambda(X \times Y) = \int_X \left\{\int_Y d\nu_x\right\}d\mu = \int_X \nu(Y)\,d\mu = \mu(X)\nu(Y)$$

for every measurable X and Y, and the distribution is independent. Conversely, if this condition is given, it follows for any given Y that

$$\int_X \nu_x(Y)\,d\mu = \lambda(X \times Y) = \mu(X)\nu(Y)$$

and therefore $\nu_x(Y) = \nu(Y)$ for almost all x.

The following theorem shows that the conditional probability measures ν_x exist, and the conditions of the last theorem hold, when $\mathscr{Y}$ (but not necessarily $\mathscr{X}$) is a real vector space $\mathscr{R}^k$.

Theorem 30. *If $\mathscr{Y} = \mathscr{R}^k$, a probability measure ν_x can be defined on $\mathscr{Y}$ for almost all x to satisfy the conditions of Theorem 29.*

It is enough to give the proof when $k = 1$. The same argument then extends without essential change, but with minor complications, to the general case. We define $\nu(I/x)$ for all simple sets I consisting of unions of finite numbers of intervals $a \leqslant y < b$ with rational end points. Since the system of all these simple sets is countable, the set of points x to be excluded in the definition of $\nu(Y/x)$ for all of them can remain a null set. If I_1, I_2 are two simple sets, we have

$$\int_X \nu(I_1 \cup I_2/x)\,d\mu = \lambda(I_1 \cup I_2 \times X) = \lambda(I_1 \times X) + \lambda(I_2 \times X)$$

$$= \int_X [\nu(I_1/x) + \nu(I_2/x)]\,d\mu$$

for every measurable X, and therefore

$$\nu(I_1 \cup I_2/x) = \nu(I_1/x) + \nu(I_2/x)$$

for almost all x. Since the finite addivity of $\nu(I/x)$ over the simple sets can be expressed by a countable number of relations of this kind, we may suppose that for almost all x, $\nu(I/x)$ is finitely additive. The complete additivity of $\nu(I/x)$ can now be expressed by a countable set of conditions of the type $\nu(I_n/x) \to \nu(I/x)$ for almost all x, where $I_n \uparrow I$. This follows for any one simple set and the sequence I_n associated with it from the fact that

$$\int_{\mathscr{X}} [\nu(I/x) - \nu(I_n/x)]\, d\mu = \lambda(\mathscr{X}, I) - \lambda(\mathscr{X} I_n) = \nu(I - I_n) = o(1) \quad \text{as} \quad n \to \infty$$

and $\nu(I/x) - \nu(I_n/x)$ is non-negative and decreasing as $n \to \infty$ for almost all x and so has limit zero by Theorem 20 of Chapter 2.

The theorem remains true for more general spaces, although we shall have no occasion to use the fact. It is sufficient that $\mathscr{Y}$ should be a topological space and that its measure should be an extension of an additive set function over a countable ring of simple sets each containing a compact set of measure arbitrarily close to its own.

After Theorem 30, the theory of dependence between variables x and y in real vector spaces can be developed in more concrete terms and it becomes useful to introduce a number of concepts and parameters which have a clear practical significance. It is enough to illustrate this in the case in which x and y are both single real numbers. The extensions to higher dimensions are more complicated but involve no basically new ideas.

We can assume then that the conditional probability ν_x is defined for almost all x and that the conditional mean $m(x)$ is defined by

$$m(x) = \mathbf{E}\{y/x\} = \int y\, d\nu_x.$$

The graph $y = m(x)$ is called the **regression curve** of y on x. It has an interesting minimal property in the cases when y has a finite variance.

Theorem 31. *Suppose that* (x, y) *has a joint distribution in which* y *has a finite variance. The conditional distribution of* y *has finite variance* $\sigma(x)$ *for almost all* x*, and the regression function* m *of* x *on* y *is the function which minimises*

$$\mathbf{E}\{[y - q(x)]^2\}$$

for all real functions q.

The existence of $\sigma(x)$ for almost all x follows from Theorem 29, from which we also deduce, by putting $\alpha(x, y) = [y - q(x)]^2$, that

$$\mathbf{E}\{[y - q(x)]^2\} = \int \left\{ \int [y - q(x)]^2 \, d\nu_x \right\} d\mu.$$

Since $\int [y - q(x)]^2 \, d\nu_x$ is obviously least, for each fixed x, when $q(x) = m(x)$, the conclusion then follows.

The number $\mathbf{E}\{[y - m(x)]^2\}$ is the mean square deviation of y from its conditional mean, and gives a measure of the deviation of the distribution from the *completely determinate* state in which y takes a single assigned value with probability 1 for every x. The **correlation ratio** θ of y on x is defined by

$$\theta^2 = 1 - \sigma^{-2} \mathbf{E}\{[y - m(x)]^2\}$$

and obviously lies between 0 and 1 and approaches the value 1 for a distribution with y concentrated near to its regression line.

The problem of choosing the function q to minimise $\mathbf{E}\{[y - q(x)]^2\}$ takes a different form if, instead of allowing a choice of $y(x)$ for all functions of x, we restrict the choice to some specified class. Among these classes the polynomials of given degree are most commonly used, and the polynomial p_r of degree r for which

$$E\{[y - p_r(x)]^2\}$$

is minimum defines the **polynomial regression curve** $y = p_r(x)$ of y on x. The **linear regression** given by the case $r = 1$ is particularly important, and it is possible in this case to carry the analysis rather further. We consider polynomials $p(x) = a + bx$ with arbitrary a and b, and find that

$$\begin{aligned} \mathbf{E}\{[y - p(x)]^2\} &= \mathbf{E}\{[y - a - bx]^2\} \\ &= \mathbf{E}\{[y - m - b(x - l) + m - bl - a]^2\}, \end{aligned}$$

where (l, m) is the mean of (x, y) for the whole distribution. This gives

$$\begin{aligned} \mathbf{E}\{[y - p(x)]^2\} = {} & \mathbf{E}\{(y - m)^2\} - 2b\mathbf{E}\{(x - l)(y - m)\} + b\mathbf{E}\{x - l)^2\} \\ & + (m - bl - a)^2 \end{aligned}$$

and this is plainly least when $b = \rho\sigma/\tau$, $a = m - l\rho\sigma/\tau$, where

$$\sigma^2 = \mathbf{E}\{(y - m)^2\}, \qquad \tau^2 = \mathbf{E}\{(x - l)^2\}$$

and ρ is the regression coefficient $E\{(x - l)(y - m)\}/\sigma\tau$. The linear regression line is therefore

$$y = p_1(x) = m + \rho\sigma(x - l)/\tau.$$

The actual minimal value of $\mathbf{E}\{[y - p(x)]^2\}$, attained when $y = p_1(x)$ is the linear regression line, has the value $\sigma^2(1 - \rho^2)$. The total variance σ^2 of y

can be expressed as the sum of two components in the form

$$\sigma^2 = \mathbf{E}\{y-m)^2\} = \mathbf{E}\{(y-m(x))^2\} + \mathbf{E}\{(m(x)-m)^2\}$$
$$= \sigma^2(1-\theta^2) + \sigma^2\theta^2,$$

where θ is the correlation ratio. The two terms on the right are the components of the variance due respectively to the deviation of y from its conditional mean and the deviation of the conditional mean from the overall mean m. The first term is therefore the residual variance which remains after as great a component as possible has been removed by a determinate function with values $y = m(x)$. A small value of this residual, given by a value of θ near to 1, indicates that the distribution is closely concentrated near a single curve $y = m(x)$. A smaller value of θ indicates that the random component is larger.

The total variance cannot be expressed in quite the same way for linear regressions, but we still have the residual variance $\sigma^2(1-\rho^2)$ when the variance has been reduced by the greatest possible amount by the subtraction from y of a linear term $a+bx$. This vanishes if and only if the whole distribution is concentrated on a line and there is a determinate linear relationship between y and x.

The condition that ρ is near to ± 1 therefore has a precise and explicit interpretation, but no such clear-cut interpretation is possible for other values. In particular, the variables are said to be **uncorrelated** when $\rho = 0$, and it is obvious that independent variables are uncorrelated but that the converse is not generally true. In fact, $\rho = 0$ means that the linear regression is parallel to the x-axis, and this is a much weaker condition than independence. Moreover, except in the cases when $|\rho|$ is near to 1, the value of ρ gives little information about the deviation of y from the determinate form. For example $\rho = 0$ if y takes the value $\cos x$ with probability 1 for every x in the interval $-\pi \leqslant x \leqslant \pi$.

6.5 Limit processes: random sequences and functions

Limit processes may arise in several different ways from sequences of random variables. In the simplest of these we are concerned only with the limiting behaviour of a sequence of distribution functions, and there is no need to consider the joint distributions of the associated random variables, although these may exist and be known.

The appropriate form of convergence is that in which the sequence of distribution functions F_n converges to a distribution function F at every continuity point of the latter, and we denote this property by writing $F_n \to F$.

The next two theorems show that this is equivalent to the convergence everywhere of the sequence of characteristic functions ϕ_n to a function ϕ continuous at 0.

Theorem 32. *If F_n, F are distribution functions with characteristic functions ϕ_n, ϕ, and if $F_n \to F$, then $\phi_n(t) \to \phi(t)$ uniformly in any finite interval.*

If $\varepsilon > 0$, we can choose an interval I and positive integer N so that

$$\int_{I'} \mathrm{d}F < \tfrac{1}{2}\varepsilon, \int_{I'} \mathrm{d}F_n < \tfrac{1}{2}\varepsilon \quad \text{for} \quad n \geqslant N.$$

Moreover, since F_n has at most a countable set of discontinuities, we may assume that F and F_n are continuous at the end points of I. Then since $|\mathrm{e}^{\mathrm{i}xt}| = 1$, it follows from Theorem 6 of Chapter 3 and Theorem 18 of Chapter 2 that for $n \geqslant N$,

$$\begin{aligned} |\phi_n(t) - \phi(t)| &\leqslant \varepsilon + \left| \int_I \mathrm{e}^{\mathrm{i}tx} \, \mathrm{d}(F_n - F) \right| \\ &\leqslant \varepsilon + |[\mathrm{e}^{\mathrm{i}tx} F_n(x) - F(x)]_I| + |t| \int_I |F_n(x) - F(x)| \, \mathrm{d}x \\ &= \varepsilon + o(1) \end{aligned}$$

Theorem 33. *If ϕ_n is the characteristic function of the distribution function F_n for $n = 1, 2, 3, \ldots$ and if $\phi_n \to \phi$ for all t, where ϕ is continuous at 0, then ϕ is the characteristic function of a distribution function F and $F_n \to F$.*

Let $F(x) = \limsup F_n(x)$ as $n \to \infty$, so that $F(x)$ increases and $0 \leqslant F(x) \leqslant 1$ in $-\infty < x < \infty$. Using Theorem 26 of Chapter 5, and the notation

$$F^{\delta}(x) = \frac{1}{2\delta} \int_{x-\delta}^{x+\delta} F(u) \, \mathrm{d}u,$$

we have

$$F_n(2D) - F_n(-2D) \geqslant F_n^D(D) - F_n^D(-D) = \frac{1}{\pi} \int \left(\frac{\sin t}{t}\right)^2 \phi_n\left(\frac{t}{D}\right) \mathrm{d}t,$$

and since $\phi_n \to \phi$ and $|\phi_n| \leqslant 1$, it follows from Theorem 18 of Chapter 2 that

$$F(2D) - F(-2D) \geqslant \frac{1}{\pi} \int \left(\frac{\sin t}{t}\right)^2 \phi\left(\frac{t}{D}\right) \mathrm{d}t$$

for any positive D. Since $\phi(t)$ is continuous at 0 and $\phi(0) = 1$, it follows again from Theorem 18 of Chapter 2 on letting $D \to \infty$ that $F(\infty) - F(-\infty) \geqslant 1$, and $F(x)$ is therefore a distribution function.

Now let α, x be any real numbers and let $\delta > 0$. It follows from Theorem 26 of Chapter 5 that $F_n^{\delta}(x) - F_n^{\delta}(\alpha)$ tends to a limit as $n \to \infty$. If

$\varepsilon > 0$, we can choose α so that $F(\alpha+\delta) < \varepsilon$, and since $\liminf F_n^\delta(\alpha) \geq 0$, it then follows that

$$\limsup F_n^\delta(x) - \liminf F_n^\delta(x) \leq \limsup F_n^\delta(\alpha) \leq F(\alpha+\delta) < \varepsilon,$$

and since this holds for every positive ε, the left-hand side must vanish and

$$\limsup F_n^\delta(x) = \liminf F_n^\delta(x).$$

Hence, since $F_n(x)$ increases with x,

$$\begin{aligned} F_n(x-2\delta) &\leq F_n^\delta(x-\delta) \leq F_n(x) \\ F(x-2\delta) &\leq \limsup F_n^\delta(x-\delta) = \liminf F_n^\delta(x-\delta) \\ &\leq \liminf F_n(x). \end{aligned}$$

This is true for every positive δ and so $F(x-0) \leq \liminf F_n(x)$ and this, together with the definition of $F(x)$, is enough to show that $F_n \to F$.

It is important to observe that the notion of a limiting *random variable* need not appear, but if we do wish to relate the behaviour of random variables x_n to such a limit variable x, it is necessary to define the joint distributions of (x, x_n) for every n. There are two important definitions based on this. We say that a random real variable x_n **converges in probability** to x, and write $x_n \to x$ in probability, if

$$\lim_{n\to\infty} \mathbf{P}\{|x_n - x| > \eta\} = 0$$

for every positive η, $\mathbf{P}$ being the joint probability in the product space of x and x_n.

It is plain that $x_n \to x$ in probability means that $x_n - x \to 0$ in probability, and this can be expressed in terms of the distribution functions of $x_n - x$. In particular, if C is a constant, $x_n \to C$ in probability is equivalent to the condition that $F_n \to D_c$, where $D_c(x) = D(x-C)$, and $D(x)$ is the singular distribution function with a single discontinuity of magnitude 1 at $x = 0$.

The second definition is of mean convergence. We say that x_n **converges in mean of order** p $(p \geq 1)$ to a random variable x if the distribution of $x_n - x$ is known from the joint distribution of (x, x_n) and if $\mathbf{E}\{|x|^p, \mathbf{E}\{|x_n|^p\}$ are finite for every n and

$$\lim_{n\to\infty} \mathbf{E}\{|x_n - x|^p\} = 0.$$

It is clear that this is equivalent to saying that $x_n - x$ converges in mean to 0. In particular, $x_n \to 0$ in mean is equivalent to

$$\lim_{n\to\infty} \int |x|^p \, dF_n = 0.$$

Before discussing the general theory of convergence of distribution functions, we mention some well known and simple special cases.

Theorem 34. (i) *If $x_n \to x$ probability, then $F_n \to F$.* (ii) *If $x_n \to x$ in mean of order p, then $x_n \to x$ in probability.*

Let a be a point of continuity of F. Let $\varepsilon > 0$ and define η so that $F(a+\eta+0) \leqslant F(a)+\varepsilon$. Then we can choose N so that $\mathbf{P}\{|x-x_n|>\eta\} \geqslant \varepsilon$ for $n \geqslant N$, and then for $n \geqslant N$,

$$\begin{aligned}\mathbf{P}\{x_n \leqslant a\}+\mathbf{P}\{x>a+\eta\} &\leqslant 1+\mathbf{P}\{x_n \leqslant a, x>a+\eta\}\\ &\leqslant 1+\mathbf{P}\{|x_n-x|>\eta\} \leqslant 1+\varepsilon,\end{aligned}$$

and

$$\begin{aligned}F_n(a) \leqslant F_n(a+0) &= \mathbf{P}\{x_n \leqslant a\}\\ &\leqslant 1-\mathbf{P}\{x>a+\eta\}+\varepsilon\\ &= \mathbf{P}\{x \leqslant a+\eta\}+\varepsilon\\ &= F(a+\eta+0)+\varepsilon \leqslant F(a)+2\varepsilon.\end{aligned}$$

A similar argument shows that $F_n(a) \geqslant F(a)-2\varepsilon$, and the conclusion (i) follows.

Part (ii) follows from the inequality

$$\mathbf{E}\{|x_n-x|^p\} \geqslant \varepsilon^p \mathbf{P}\{|x_n-x|>\varepsilon\}.$$

Theorem 35 (Tchebycheff). *If x_n has mean m_n and standard deviation σ_n, and if $\sigma_n \to 0$, then $x_n - m_n \to 0$ in probability.*

This follows directly from the case $p=2$ of Theorem 34.

Theorem 36 (Bernoulli's weak law of large numbers). *Let $\xi_1, \xi_2, \ldots$ be independent random real numbers with standard deviations $\sigma_1, \sigma_2, \ldots$ and let*

$$x_n = \frac{1}{n}\sum_{v=1}^{n} \xi_v, \qquad m_n = \mathbf{E}\{x_n\} = \sum_{v=1}^{n} \mathbf{E}\{\xi_v\}.$$

Then $x_n - m_n \to 0$ in probability if $\sum_{v=1}^{n} \sigma_v^2 = o(n^2)$.

The conditions ensure that $x_n - m_n$ has mean 0 and standard deviation tending to 0, and the conclusion then follows from Tchebycheff's inequality (Theorem 4).

Theorem 37 (Khintchine). *If $\xi_1, \xi_2, \ldots$ are independent random real numbers with the same distribution and finite mean m, then*

$$x_n = \frac{1}{n}\sum_{v=1}^{n} \xi_v \to m$$

in probability.

(We notice first that this follows from Theorem 36 if ξ_v have finite variances, but not in the more general case in which this is not assumed.)

Let ϕ, F be the characteristic and distribution functions of ξ_v, so that the characteristic function of x_n has value $[\phi(t/n)]^n$ and

$$\phi(t) = 1 + itm + \int (e^{itx} - 1 - itx)\, dF.$$

Since $|(e^{itx} - 1 - itx)/t| \leqslant 2|x|$ for all t, and tends to 0 for for every fixed x, as $t \to 0$, it follows from the Lebesgue convergence theorem (Theorem 18 of Chapter 2) that $\phi(t) = 1 + itm + o(t)$ and that

$$\left[\phi\left(\frac{t}{n}\right)\right]^n = \left[1 + \frac{itm}{n} + o\left(\frac{t}{n}\right)\right]^n \to e^{itm}$$

for every t as $n \to \infty$. Since e^{itm} is the characteristic function $D(x-m)$, the conclusion follows.

Theorem 38 (Poisson). *Let F_n be the distribution function of a random integer x_n with a binomial distribution given by*

$$\mathbf{P}\{x_n = v\} = \binom{n}{v} p_n^v q_n^{n-v},$$

where $p_n = c/n$, $q_n = 1 - p_n$, and c is a positive constant. Then $F_n \to F$, where F is the Poisson distribution function with mean c.

The characteristic function of F_n is given by

$$\phi_n(t) = (q_n + p_n e^{it})^n = \left(1 + \frac{c(e^{it} - 1)}{n}\right)^n$$
$$= e^{c(e^{it}-1)} + o(1),$$

as $n \to \infty$, which is sufficient since this is the Poisson characteristic function.

Theorem 39 (de Moivre). *If $\xi_1, \xi_2, \ldots$ are independent random real numbers with the same distribution and characteristic functions F and ϕ, with mean 0 and finite standard deviation σ, then the distribution function F_n of*

$$x_n = n^{-1/2} \sum_{v=1}^{n} \xi_v$$

tends to the normal $(0, \sigma)$ distribution.

The characteristic function of x_n is defined by

$$\phi_n(t) = [\phi(tn^{-1/2})]^n,$$

where

$$\phi(t) = 1 - \tfrac{1}{2}t^2\sigma^2 + \int [\mathrm{e}^{\mathrm{i}tx} - 1 - \mathrm{i}tx + \tfrac{1}{2}t^2x^2]\,\mathrm{d}F.$$

It follows from the elementary inequality

$$\theta(u) = |\mathrm{e}^{\mathrm{i}u} - 1 - \mathrm{i}u + \tfrac{1}{2}u^2| \leqslant \min(u^2, |u^3|)$$

for real u that $|\theta(tx)/t^2| \leqslant x^2$ for all t and tends to 0, for every x, as $t \to 0$. It then follows from Lebesgue's convergence theorem that

$$\phi_n(t) = [1 - \tfrac{1}{2}t^2\sigma^2/n + o(t^2/n)]^n \to \mathrm{e}^{-(1/2)t^2\sigma^2},$$

as we require.

In each of the last three theorems, the random variable x_n can be regarded as the sum of a large number of small independent components, and the conclusion in each case is that the distribution function F_n of x_n approximates to one of the three forms—singular, Poisson, normal. The problem of generalizing these results and bringing them into a comprehensive theory constitutes the **central limit problem.** This has been carried to a very elegant and satisfactory conclusion through the work of Kolmogoroff, Khintchine, Levy and others. Here we have space only for the restricted but still very important and useful case in which the variances of the random variables involved are all finite and bounded.

Theorem 40 (Liapounoff). *Suppose that, for every positive integer n, x_{nv} are independent random real numbers with mean 0, standard deviations σ_{nv} and distribution function F_{nv}. Suppose also that $x_n = \sum_v x_{nv}$ has distribution function F_n, that $\sum \sigma_{nv}^2 = 1$ and*

$$L_n(\eta) = \sum_v l_{nv}(\eta),\ l_{nv}(\eta) = \int_{|x| \geqslant \eta} x^2\,\mathrm{d}F_{nv}$$

for $\eta \geqslant 0$.

(i) *If, for every $\eta > 0$, $L_n(\eta) \to 0$ as $n \to \infty$, then F_n tends to the normal* (0, 1) *distribution.*
(ii) *Conversely, if $\sigma_{nv} \to 0$ uniformly in v as $n \to \infty$, and if F_n tends to normal* (0, 1), *then $L_n(\eta) \to 0$ as $n \to \infty$.*

We note first that

$$\max_v \sigma_{nv}^2 \leqslant \max_v \int_{|x|<\eta} x^2\,\mathrm{d}F_{nv} + \max_v \int_{|x| \geqslant \eta} x^2\,\mathrm{d}F_{nv} \leqslant \eta^2 + L_n(\eta)$$

for $\eta > 0$ and it follows that the conditions of (i) imply that $\sigma_{nv} \to 0$ uniformly in v as $n \to \infty$ and that this can be assumed in both parts of the theorem. Then

$$\phi_{nv}(t) = \int \mathrm{e}^{\mathrm{i}tx}\,\mathrm{d}F_{nv} = 1 - \tfrac{1}{2}t^2\sigma_{nv}^2 + \tfrac{1}{2}t^2 l_{nv} + \alpha_{nv} + \beta_{nv},$$

where for $\eta > 0$

$$0 \leqslant l_{nv} \leqslant \sigma_{nv}^2,$$

$$\alpha_{nv} = \int_{|x| \geqslant \eta} (e^{itx} - 1 - itx)\, dF_{nv},$$

$$\beta_{nv} = \int_{|x| < \eta} (e^{itx} - 1 - itx + \tfrac{1}{2}t^2x^2)\, dF_{nv}.$$

Using the elementary inequalities

$$|e^{iu} - 1 - iu| \leqslant 2\,|u|,\ |e^{iu} - 1 - iu + \tfrac{1}{2}u^2| \leqslant |u|^3$$

for all real u, we have

$$|\alpha_{nv}| \leqslant |t| \int_{|x| \geqslant \eta} |x|\, dF_{nv} \leqslant 2\eta\, |t|\, l_{nv} \leqslant 2\eta\, |t|\, \sigma_{nv}^2, \qquad \sum |\alpha_{nv}| \leqslant 2\eta\, |t|, \quad (1)$$

$$|\beta_{nv}| \leqslant |t|^3\, \eta \int_{|x| < \eta} x^2\, dF_{nv} \leqslant \eta\, |t|^3\, \sigma_{nv}^2, \qquad \sum |\beta_{nv}| \leqslant \eta\, |t|^3, \quad (2)$$

and σ_{nv}^2, l_{nv}, α_{nv}, β_{nv} all tend to 0 uniformly in v as $n \to \infty$. Since $\log(1+z)$ is differentiable at $z = 0$, $\log(1+z) - z = o(|z|)$ as $z \to 0$ and so $\log \phi_{nv}(t) + \frac{1}{2}t^2\sigma_{nv}^2 - t^2 l_{nv} - \alpha_{nv} - \beta_{nv} = o(\sigma_{nv}^2)$ uniformly in v as $n \to \infty$. Since $\phi_n(t) = \prod_v \phi_{nv}(t)$ by Theorem 11, and $\sum \sigma_{nv}^2 = 1$, $\sum l_n = L$,

$$\begin{aligned} |\log \phi_n(t) + \tfrac{1}{2}t^2 - \tfrac{1}{2}t^2 L_n(\eta)| &\leqslant \sum |\alpha_{nv}| + \sum |\beta_{nv}| + o(1) \\ &\leqslant \eta\, |t|\, (2 + t^2) + o(1) \end{aligned}$$

by (1) and (2).

In case (i), if we keep t and $\eta > 0$ fixed and let $n \to \infty$, we have $L_n(\eta) \to 0$

$$\limsup |\log \phi_n(t) + \tfrac{1}{2}t^2| \leqslant \eta\, |t|\, (2 + t^2),$$

and since this holds for all positive η, we have $\log \phi_n(t) \to -\frac{1}{2}t^2$ as required. In case (ii), $\log \phi_n(t) + \frac{1}{2}t^2 \to 0$, as $n \to \infty$,

$$\limsup \tfrac{1}{2}t^2 L_n(\eta) \leqslant \eta\, |t|\, (2 + t^2).$$

But the left-hand side is positive and non-increasing for $\eta > 0$ and the inequality implies that it must be 0 for all $\eta > 0$.

It is interesting to note that the random variable x_n in Theorem 38 is the sum of small random variables x_{nv} (with distribution functions F_{nv} which are constant except for discontinuities $1 - c/n$ at 0 and c/n at 1) while x_n does not tend to normal. The explanation is simply that the Poisson distributions F_{nv} give $L(\eta) = c$ for $0 < \eta < 1$ and the condition of Theorem 40(i) is not satisfied. On the other hand, Theorem 40(ii) is not applicable or relevant to the case in which x_{nv} are themselves normal, for the distribution of x_n and the limiting distribution are normal by Theorem 15(iii) and it does not follow that either $L_n(\eta) \to 0$ or $\sigma_{nv} \to 0$ uniformly in v.

The limit processes so far described in this section do not involve the concept of a random sequence or function, but we have already seen in Section 6.3 that the idea of a random vector $x = (x_1, x_2, \ldots, x_n)$ can be derived from that of probability measure in $\mathcal{R}^n$. If we think of the vector as a real-valued function defined over a finite set of integers, it is easy to see how the idea of a random sequence or function can be developed. We suppose that $\mathcal{X}$ is the space of all real-valued functions $x(t)$ defined over an arbitrary index space $\mathcal{T}$, and say that $x(t)$ is a **random function** in $\mathcal{X}$ if a probability measure $\mu(X)$ is defined in $\mathcal{X}$ and the statement that $\mathbf{P}\{x \in X\} = p$ is interpreted to mean that X is measurable with respect to μ and $\mu(X) = p$. This is quite analogous to the idea of a random real number, and we get a **random** n**-vector** if $\mathcal{T}$ is a finite space and a **random sequence** if $\mathcal{T}$ is the countable set of positive integers. In the last case, it is usually convenient to use the conventional notation $x = x_1, x_2, \ldots = \{x_v\}$. Otherwise, there is no need to put any restriction on $\mathcal{T}$, but in the illustration we consider it will always be a subset of the real line. In many applications it is useful to use t to denote time, and the random function is then called a **random** or **stochastic process** or (particularly when the variable is integral) a **time series**.

The basic problem can now be stated in a form analogous to that associated with the idea of a random variable in its simplest form. It is to establish the existence of a probability measure in $\mathcal{X}$ to have certain properties, which are usually to ensure that a certain specified family of sets are measurable and have specified measures. These specified measures must, of course, be mutually consistent. The specified sets which turn out to be appropriate are the simple sets defined as finite unions of rectangular sets of functions $x(t)$ satisfying a finite set of conditions of the type $a_v \leq x(t_v) < b_v$.

It is plain that the simple sets associated with a particular finite system of points t_v form a ring, and that the totality of all such simple sets also forms a ring. The basic theorem of Kolmogoroff can be stated in a form closely analogous to Theorem 24 of Chapter 2.

Theorem 41 (Kolmogoroff). *Suppose that a non-negative additive set function $\mu_0(1)$ is given on the ring of all simple sets I of $\mathcal{X}$ and that for every finite set of points $(t_1, t_2, \ldots, t_n)$, the values of $\mu_0(I)$ on the ring of sets associated with $t_1, \ldots, t_n$ are those of a probability distribution in $\mathcal{R}^n$. Then it is possible to define a probability measure $\mu(X)$ in $\mathcal{X}$ in such a way that every simple set I is measurable and $\mu(I) = \mu_0(I)$.*

After Theorem 2 of Chapter 1 and Theorem 24 of Chapter 2, it is sufficient to show that if I_n are simple and $I_n \downarrow 0$, then $\mu(I_n) \to 0$. We suppose the contrary and derive a contradiction. Since the number of values of t associated with any one I_n is finite, the set of all such points t is countable, and they can be arranged as a sequence (t_i). Now each I_n is the union of a finite set of rectangular sets and we can select one of these for $n = 1, 2, \ldots,$ so that it contains a *closed* rectangular set J_n with the property that $\lim_{m\to\infty} \mu(J_n \cap I_m) > 0$. Also, we may choose J_n so that $J_{n+1} \subset J_n$, and we have therefore a decreasing sequence of closed non-empty rectangular sets J_n defined by

$$a_{i_n} \leqslant y_i \leqslant b_{i_n} \qquad (i = 1, 2, \ldots, i_n).$$

For each i there is at least one point y_i which is contained in all the intervals $[a_{i_n}, b_{i_n}]$, and any function $x(t)$ for which $x(t_i) = y_i$ for $i = 1, 2, 3, \ldots$ belongs to every J_n and so to every I_n. This is impossible since $I_n \downarrow 0$, and therefore we have $\mu(I_n) \to 0$.

The theorem applies immediately to the definition of random sequences when $\mathcal{T}$ is the space of positive integers. The case in which the terms x_n are independent is of particular interest and can be formulated more simple as follows.

Theorem 42. *Suppose that $F_1, F_2, \ldots$ are distribution functions. Then it is possible to define a probability measure in the space of real sequences $x_1, x_2, \ldots$ in such a way that the set of sequences satisfying conditions*

$$a_i \leqslant x_{n_i} < b_i$$

for any finite set of positive integers n_i is measurable and has measure

$$\prod [F(b_i - 0) - F(a_i - 0)].$$

More generally, if μ_v is the probability distribution of x_v (defined by F_v), the set of sequences x_n satisfying $x_{n_i} \in X_i$ for any finite set of integers n_i and measurable sets X_i of $\mathcal{R}$ is measurable and has measure $\prod \mu_{n_i}(X_i)$.

The random function defined by this theorem is said to have *independent* terms x_v with distributions F_v. The theorem provides a completely satisfactory basis for the discussion of the probability theory of independent sequences, and their properties of boundedness, convergence, summability, and so on which can be expressed in terms of countable sets of conditions. The theorems which follow serve to illustrate some of the methods which may be used. The first is a general result of wide application, called the 0 or 1 principle of Borel.

Theorem 43. *The probability that a random sequence of independent variables should have a property (e.g. convergence) which is not affected by changes in a finite number of terms is equal to* 0 *or* 1.

Let X be the set of sequences with the given property, so that our hypothesis is that for every $N \geqslant 1$,

$$X = \mathcal{R}_1 \times \mathcal{R}_2 \times \ldots \times \mathcal{R}_N \times X_N,$$

where X_N is a set in the space of sequences $(x_{n+1}, x_{n+2}, \ldots)$. If I is any simple set in $\mathcal{X}$, it follows that $I = I \cap X_N$ for large enough N and

$$\mu(I \cap X) = \mu(I)\mu(X_N) = \mu(I)\mu(X).$$

Since this holds for all simple sets, it extends to all measurable sets Y and $\mu(Y \cap X) = \mu(Y)\mu(X)$. In particular, putting $Y = X$,

$$\mu(X) = [\mu(X)]^2, \ \mu(x) = 0 \quad \text{or} \quad 1.$$

We can now proceed to discuss some general questions on the convergence of series of independent random real variables. After the last theorem, the series converges with probability 0 or 1. If the value is 1, the sum of the series is a function of x on $\mathcal{X}$ and is therefore itself a random variable. The main problem is to determine how the convergence or divergence of the series is related to the sequence of distribution or characteristic functions of the variables x_v.

The following theorem gives a completely satisfactory answer to one direction.

Theorem 4.4. *If $x_1, x_2, \ldots$ are independent random real numbers, and $s_n = \sum_{v=1}^{n} x_v$ converges to the (random) sum s with probability 1, then $s_n \to s$ in probability and the distribution and characteristic functions of s_n tend to those of s.*

If we think of s_n and s as functions of x over $\mathcal{X}$, and if $\varepsilon > 0$, we have

$$\mathbf{P}\{|s_n - s| > \varepsilon\} \leqslant \mathbf{P}\left\{\sup_{v \geqslant n} |s_v - s| > \varepsilon\right\} = \mu(X_n),$$

where μ is the probability distribution of x in $\mathcal{X}$ and X_n is the set of series for which $\sup_{v \geqslant n} |s_v - s| > \varepsilon$. These sets X_n decrease and tend to the null set of divergent series, and therefore $\mu(X_n) \to 0$ for every positive ε, and $s_n \to s$ in probability. The second part follows at once from Theorem 34.

The next theorem is the basis of the converse theory in which we deduce convergence with probability 1 from properties of the distribution. It is restricted to series of terms with finite variance, but this is not a serious limitation to the application of the theorem.

Theorem 45 (Kolmogoroff's inequality). *If x_v are independent, with means*

0 *and standard deviations* σ_v, *and if*

$$s_n = \sum_{v=1}^{n} x_v, \qquad t_n = \sup_{v \leqslant n} |s_v|, \qquad \varepsilon > 0,$$

then

$$\mathbf{P}\{t_n \geqslant \varepsilon\} \leqslant \varepsilon^{-2} \sum_{v=1}^{n} \sigma_v^2.$$

The sets $X_v = \{x: |s_v| \geqslant \varepsilon, t_{v-1} < \varepsilon\}$ are disjoint for $v = 1, 2, \ldots$ and

$$X = \{x: t_n \geqslant \varepsilon\} = \bigcup_{v=1}^{n} X_v.$$

Moreover, since x_v are independent,

$$\begin{aligned}\sum_{v=1}^{n} \sigma_v^2 = \int s_n^2 \, d\mu \geqslant \int_X s_n^2 \, d\mu &= \sum_{v=1}^{n} \int_{X_v} s_n^2 \, d\mu \\ &= \sum_{v=1}^{n} \int_{X_v} (s_v + x_{v+1} + x_{v+2} + \ldots + x_n)^2 \, d\mu \\ &= \sum_{v=1}^{n} \int_{X_v} s_v^2 \, d\mu + \sum_{v=1}^{n-1} \mu(X_v) \sum_{i=v+1}^{n} \sigma_i^2,\end{aligned}$$

since X_v is the product of the whole of the space of sequences $(x_{v+1}, x_{v+2}, \ldots)$ and a set of measure $\mu(X_v)$ in the finite dimensional space of $(x_1, x_2, \ldots, x_v)$. Hence

$$\sum_{v=1}^{n} \sigma_v^2 \geqslant \sum_{v=1}^{n} \int_{X_v} s_v^2 \, d\mu \geqslant \varepsilon^2 \sum_{v=1}^{n} \mu(X_v) = \varepsilon^2 \mu(X),$$

as we require.

The simplest form of converse of Theorem 44 follows easily from this.

Theorem 46. *If* x_v *are independent with mean* m_v *and standard deviations* σ_v, *and if* $\sum \sigma_v^2 < \infty$, *then* $\sum (x_v - m_v)$ *converges with probability* 1.

It is obviously sufficient to prove the theorem in the case $m_v = 0$, and then it follows from Theorem 45 applied to variables $x_{m+1}, x_{m+2}, \ldots, x_{m+n}$ that

$$\mathbf{P}\Big\{\sup_{1 \leqslant v \leqslant n} |s_{m+v} - s_m| > 1/k\Big\} \leqslant k^2 \sum_{v=m+1}^{m+n} \sigma_v^2,$$

and since $|s_{m+v} - s_{m+q}| > 1/k$ with $1 \leqslant q \leqslant v \leqslant n$ implies that $|s_{m+v} - s_m| > 1/2k$ or $|s_{m+q} - s_m| > 1/2k$, we have

$$\mathbf{P}\Big\{\sup_{1 \leqslant q \leqslant v \leqslant n} |s_{m+v} - s_{m+q}| > 1/k\Big\} \leqslant 2k^2 \sum_{v=m+1}^{m+n} \sigma_v^2.$$

Therefore

$$\mathbf{P}\left\{\lim_{m\to\infty}\sup_{1\leqslant q\leqslant v}|s_{m+v}-s_{m+q}|>0\right\}=\lim_{k\to\infty}\mathbf{P}\left\{\lim_{m\to\infty}\sup_{1\leqslant q\leqslant v}|s_{m+v}-s_{m+q}|>1/k\right\}$$

$$=\lim_{k\to\infty}\lim_{m\to\infty}\lim_{n\to\infty}\mathbf{P}\left\{\sup_{1\leqslant q\leqslant v\leqslant n}|s_{m+v}-s_{m+q}|>1/k\right\}$$

$$\leqslant\lim_{m\to\infty}2k^2\sum_{v=m+1}^{\infty}\sigma_v^2=0$$

by the convergence of $\sum\sigma_v^2$. But

$$\lim_{m\to\infty}\sup_{1\leqslant q\leqslant v}|s_{m+v}-s_{m+q}|>0$$

is the necessary and sufficient condition for the divergence of $\sum x_v$, by the general principle of convergence, and the conclusion follows.

The extension of the idea of a random sequence to a random function defined on the whole real line involves problems of quite a different order of difficulty and we do no more than describe some of the basic concepts. The foundation of the theory is Kolmogoroff's existence theorem (Theorem 41) which shows that a significant measure can be defined in $\mathscr{X}$ in such a way that all the simple sets are measurable and have any assigned measures, provided that these are mutually consistent. The measure defined in this way over the minimal Borel extension of the simple sets may be called a ***K*-measure** and it is determined and characterised by the totality of all the finite dimensional distributions associated with all finite sets of points in $\mathscr{T}$. In fact, we may call these the finite dimensional **generating distributions** and speak of the K-measure defined by them.

The first major problem in the theory arises from the fact that a K-measure is generally not extensive enough, in that it leaves as unmeasurable many important classes of functions $x(t)$. For example, we find that the class of continuous functions is not measurable with respect to a K-measure. The reason for this is that, unlike the convergence properties discussed in the first part of the section, continuity, differentiability, boundedness and other properties cannot be expressed in terms of a *countable* number of conditions of the kind used in the specification of the simple sets.

In order to deal with these properties, it is therefore necessary to introduce *extensions* of a K-measure. It is generally possible to extend a K-measure in many different ways. Some of the extensions may include or be extensions of others, while some systems of extensions may be mutually inconsistent. But all of them must, of course, be consistent with the K-measure on which they are based.

The general problem is therefore to investigate the system of extensions of a K-measure, to determine whether there is any one of them which makes the required sets measurable and, if there are several, to choose the most useful or appropriate one. For example, in many important cases it is possible to extend the same K-measure in two different and mutually inconsistent ways so that, in the first, almost all functions are continuous while, in the second, almost all functions are discontinuous. We should normally choose the first, and the question of the existence of such an extension of a given K-measure is an important problem.

While the definition of random functions in terms of K-measures and their extensions is the most natural, it is possible, and sometimes very convenient, to use a different approach. This is to regard a random function x as a function of a random variable ω in some probability space Ω, and this is particularly convenient if it turns out that Ω can be taken to be one of the familiar spaces such as the unit interval.

For example, we get a random function in this way be defining

$$x(t) = x(\omega, t) = \omega_1 \cos t + \omega_2 \cos 2t + \ldots + \omega_n \cos nt,$$

where $\omega = (\omega_1, \omega_2, \ldots, \omega_n)$ is a random real n-vector. A random function defined in this way for all t will define a unique K-measure, and will obviously be an extension of it. But it will not be the only extension and may not necessarily be the most useful one.

In the last resort, a random function can always be treated in this way by means of the identity function over the function space $\mathscr{X}$ itself with the measure already defined in it. Apart from convenience of notation, there seems to be little advantage in this.

The classification of random functions by the properties of their measures is one of the basic problems of the theory, and we can go no further here than to introduce some of the ideas on which useful classifications have been made. It is important in all cases to recognise whether the property of the measure can be expressed as a property of its underlying K-measure or not.

First, a random function is called **stationary** if the transformation of $\mathscr{X}$ onto itself defined by the translation $x(t) \to x(t+a)$ is measure-preserving in the sense that any measurable set is transformed into a measurable set of the same measure. The property is plainly related to the measure itself, and not merely to its K-measure. Thus a K-measure may be stationary, while an extension of it is not.

A random function is said to have **independent increments** if the random numbers $x(t_i') - x(t_i)$ are independent for every finite set of non-overlapping intervals (t_i, t_i'). This property also relates only to the K-measure and its generating distributions.

A random function is said to be an L_2 **function** if, for every fixed t, $x(t)$ is an L_2 function over $\mathscr{X}$ with respect to its probability measure in $\mathscr{X}$. The whole process can then be described as a trajectory or curve, with parameter t, in the Hilbert space $L_2(\mathscr{X})$, and many of its properties can be described in terms of the **auto-correlation function**:

$$r(s,t) = \overline{r(t,s)} = \mathbf{E}\{[x(s)-m(s)][\overline{x(t)-m(t)}]\}$$
$$= \mathbf{E}\{x(s)\overline{x(t)}\} - m(s)\overline{m(t)},$$

where $m(t) = \mathbf{E}\{x(t)\}$ is the mean over $\mathscr{X}$ of the value of the function at t. Since the mean $m(t)$ is defined for all t, we can consider the random function $x(t)-m(t)$, which has mean 0 for every t, and is said to be *centred*. A particularly important type of L_2 function is the **Gaussian**, for which the generating distributions are all normal and are defined completely by the autocorrelation function.

A random function whose increments over non-overlapping intervals are uncorrelated is said to have **orthogonal increments** or to be an **orthogonal L_2 random function**. The necessary and sufficient condition for this when the function is centred is that

$$\mathbf{E}\{[x(t')-x(t)][\overline{x(s')-x(s)}]\} = 0$$

when the intervals (t, t') and (s, s') do not overlap. This can be put in the form

$$\mathbf{E}\{|[x(t')-x(t)]+[x(s')-x(s)]|^2\} = \mathbf{E}\{|x(t')-x(t)|^2\} + \mathbf{E}\{|x(s')-x(s)|^2\},$$

which means that

$$\mathbf{E}\{|x(t')-x(t)|^2\} = F(t') - F(t)$$

for some non-decreasing function F and every t, t'.

The autocorrelation function defines a weaker form of stationarity. We say that $x(t)$ is **stationary in the L_2-sense** if $r(s,t)$ depends only on $t-s$, and this implies that $r(t, t+h) = p(h)$ and $F(t+h)-F(t) = h\sigma^2$ for some positive constant σ.

It is plain that the properties of L_2 functions which have been mentioned and all other properties that can be expressed in terms of its autocorrelation function, are properties of its K-measure and are equally true of any extension of it.

Apart from the problem associated with the classification of random functions and the study of special types, there are two more fundamental problems. First, the *separability problem* is that of expressing properties of random functions in terms of countable sets of conditions. This is desirable since the presentation of measurability and probability conditions is generally limited to countable sequences of operations.

The second problem is that of determining conditions under which a random function x, regarded as a real-valued function on the product space $\mathscr{X} \times \mathscr{T}$, is measurable with respect to the product of the probability measure in X and Lebesgue or Stieltjes measure in $\mathscr{T}$. It is plainly essential to establish measurability in this sense in order to develop a natural theory of integration to embrace both integration on $\mathscr{X}$, as in the L_2 theory mentioned above, and the integration in $\mathscr{T}$ of individual or sample functions x.

For a general discussion of these and other problems an interested reader should refer to Doob† and the periodical literature.

† Doob, J. L. (1953) *Stochastic processes.* Wiley, New York.

Index